T0332617

Computer Algebra with LISP and REDUCE

Mathematics and Its Applications

Volume 72

Computer Algebra with LISP and REDUCE

An Introduction to Computer-aided Pure Mathematics

by

F. Brackx
Department of Mathematics, University of Ghent, Ghent, Belgium

and

D. Constales
Belgian National Fund for Scientific Research (NFWO), Belgium

KLUWER ACADEMIC PUBLISHERS
DORDRECHT / BOSTON / LONDON

Library of Congress Cataloging-in-Publication Data

Brackx, F.
    Computer algebra with LISP and REDUCE : an introduction to
  computer-aided pure mathematics / by F. Brackx and D. Constales.
        p.   cm. -- (Mathematics and its applications ; v. 72)
    Includes bibliographical references and index.
    ISBN 0-7923-1441-7 (HB, printed on acid free paper)
    1. Algebra--Data processing. 2. LISP (Computer program language)
  3. REDUCE (Computer program)   I. Constales, D. II. Title.
  III. Series: Mathematics and its applications (Kluwer Academic
  Publishers) ; v.72.
  QA155.7.E4B72 1991
  512'.0285--dc20   91-31211

ISBN 0-7923-1441-7

Published by Kluwer Academic Publishers,
P.O. Box 17, 3300 AA Dordrecht, The Netherlands.

Kluwer Academic Publishers incorporates
the publishing programmes of
D. Reidel, Martinus Nijhoff, Dr W. Junk and MTP Press.

Sold and distributed in the U.S.A. and Canada
by Kluwer Academic Publishers,
101 Philip Drive, Norwell, MA 02061, U.S.A.

In all other countries, sold and distributed
by Kluwer Academic Publishers Group,
P.O. Box 322, 3300 AH Dordrecht, The Netherlands.

Printed on acid-free paper

Printed in the Netherlands

'Et moi, ..., si j'avait su comment en revenir,
je n'y serais point allé.'

Jules Verne

The series is divergent; therefore we may be
able to do something with it.

O. Heaviside

One service mathematics has rendered the
human race. It has put common sense back
where it belongs, on the topmost shelf next
to the dusty canister labelled 'discarded non-
sense'.

Eric T. Bell

Mathematics is a tool for thought. A highly necessary tool in a world where both feedback and non-linearities abound. Similarly, all kinds of parts of mathematics serve as tools for other parts and for other sciences.

Applying a simple rewriting rule to the quote on the right above one finds such statements as: 'One service topology has rendered mathematical physics ...'; 'One service logic has rendered computer science ...'; 'One service category theory has rendered mathematics ...'. All arguably true. And all statements obtainable this way form part of the raison d'être of this series.

This series, *Mathematics and Its Applications*, started in 1977. Now that over one hundred volumes have appeared it seems opportune to reexamine its scope. At the time I wrote

"Growing specialization and diversification have brought a host of monographs and textbooks on increasingly specialized topics. However, the 'tree' of knowledge of mathematics and related fields does not grow only by putting forth new branches. It also happens, quite often in fact, that branches which were thought to be completely disparate are suddenly seen to be related. Further, the kind and level of sophistication of mathematics applied in various sciences has changed drastically in recent years: measure theory is used (non-trivially) in regional and theoretical economics; algebraic geometry interacts with physics; the Minkowsky lemma, coding theory and the structure of water meet one another in packing and covering theory; quantum fields, crystal defects and mathematical programming profit from homotopy theory; Lie algebras are relevant to filtering; and prediction and electrical engineering can use Stein spaces. And in addition to this there are such new emerging subdisciplines as 'experimental mathematics', 'CFD', 'completely integrable systems', 'chaos, synergetics and large-scale order', which are almost impossible to fit into the existing classification schemes. They draw upon widely different sections of mathematics."

By and large, all this still applies today. It is still true that at first sight mathematics seems rather fragmented and that to find, see, and exploit the deeper underlying interrelations more effort is needed and so are books that can help mathematicians and scientists do so. Accordingly MIA will continue to try to make such books available.

If anything, the description I gave in 1977 is now an understatement. To the examples of interaction areas one should add string theory where Riemann surfaces, algebraic geometry, modular functions, knots, quantum field theory, Kac-Moody algebras, monstrous moonshine (and more) all come together. And to the examples of things which can be usefully applied let me add the topic 'finite geometry'; a combination of words which sounds like it might not even exist, let alone be applicable. And yet it is being applied: to statistics via designs, to radar/sonar detection arrays (via finite projective planes), and to bus connections of VLSI chips (via difference sets). There seems to be no part of (so-called pure) mathematics that is not in immediate danger of being applied. And, accordingly, the applied mathematician needs to be aware of much more. Besides analysis and numerics, the traditional workhorses, he may need all kinds of combinatorics, algebra, probability, and so on.

In addition, the applied scientist needs to cope increasingly with the nonlinear world and the extra mathematical sophistication that this requires. For that is where the rewards are. Linear models are honest and a bit sad and depressing: proportional efforts and results. It is in the non-linear world that infinitesimal inputs may result in macroscopic outputs (or vice versa). To appreciate what I am hinting at: if electronics were linear we would have no fun with transistors and computers; we would have no TV; in fact you would not be reading these lines.

There is also no safety in ignoring such outlandish things as nonstandard analysis, superspace and anticommuting integration, p-adic and ultrametric space. All three have applications in both electrical engineering and physics. Once, complex numbers were equally outlandish, but they frequently proved the shortest path between 'real' results. Similarly, the first two topics named have already provided a number of 'wormhole' paths. There is no telling where all this is leading - fortunately.

Thus the original scope of the series, which for various (sound) reasons now comprises five subseries: white (Japan), yellow (China), red (USSR), blue (Eastern Europe), and green (everything else), still applies. It has been enlarged a bit to include books treating of the tools from one subdiscipline which are used in others. Thus the series still aims at books dealing with:

- a central concept which plays an important role in several different mathematical and/or scientific specialization areas;
- new applications of the results and ideas from one area of scientific endeavour into another;
- influences which the results, problems and concepts of one field of enquiry have, and have had, on the development of another.

In spite of its name, a computer is a data-processing device and not necessarily a calculating machine. One aspect of its scientific uses that is rapidly attracting more and more attention is the use of these machines for symbolic rather than numerical calculation.

For this kind of use, the programming LISP and general purpose computer algebra systems, of which REDUCE is one of the well-known ones, offer critical advantages when compared to classical high-level programming languages such as FORTRAN, PASCAL, and C.

This volume, based on a set of ERASMUS courses, offers a selfcontained, and accessible treatment of LISP and REDUCE for symbolic manipulations.

The shortest path between two truths in the real domain passes through the complex domain.	Never lend books, for no one ever returns them; the only books I have in my library are books that other folk have lent me.
J. Hadamard	Anatole France
La physique ne nous donne pas seulement l'occasion de résoudre des problèmes ... elle nous fait pressentir la solution.	The function of an expert is not to be more right than other people, but to be wrong for more sophisticated reasons.
H. Poincaré	David Butler

Amsterdam, August 1991   Michiel Hazewinkel

Contents

Preface

This text is based upon a series of lectures on LISP and REDUCE held at the Universities of Aveiro and Coimbra (Portugal), as part of the European Community's ERASMUS Project, and at the University of Gent (Belgium). It has been adapted to the newest release of REDUCE, version 3.4.

In Chapter 1 we introduce LISP and computer algebra systems in the context of computer programming and demonstrate their critical advantages when compared to classical high-level programming languages such as FORTRAN, PASCAL and C. We also discuss the price that has to be paid for these advantages, and justify our choice of LISP and REDUCE.

Chapter 2 presents LISP as a safe and interactive way of creating and manipulating data structures. We indicate how memory is organized by a LISP environment and how the elementary structural operations of LISP give full control over it to the user. Next, we survey the built-in functions of Standard LISP, starting with those common to all LISPs and ending with the more specific ones. Finally, we indicate how RLISP relates to Standard LISP and what notational advantages it offers.

Chapter 3 presents REDUCE's algebraic mode—with the occasional peek at what happens at the LISP level.

Chapters 4 and 5, finally, offer several non-trivial applications of REDUCE and RLISP.

A selected bibliography is appended.

The book is aimed at a wide audience and only presupposes a practical knowledge of algebra and calculus (as taught during the first two years at University) and, when we compare LISP and REDUCE to other programming languages, some acquaintance with FORTRAN, PASCAL or C language.

Both authors wish to thank the Belgian National Fund for Scientific Research NFWO for its essential support of the CAGe research group ('Computer Algebra at the University of Gent'). The second author is also indebted to the NFWO for supporting him as a Senior Research Assistant. Thanks are due also to Dr. H. Serras, for his drawing skills, to Dr. K. Coolsaet, for his careful reading of the manuscript, and—last but not least—to Prof. R. Delanghe, for his untiring stimulation and encouragement of our work.

Gent, September 1991.

Chapter 1

INTRODUCTION

1.1 Computers, mathematics and computer algebra

Historically, the digital computer's first applications were of a scientific nature: simulation, optimization, operational research, numerical analysis. Research in these areas was greatly stimulated by the introduction of digital computers because these could perform elementary operations on integers with limited range and on approximated real numbers, at a speed far exceeding that of a human computer.

With time, large companies and government saw that computers could help them automate certain administrative tasks, thereby reducing their cost and rationalizing them. In such applications, the emphasis no longer lay on the purely computational aspects of digital computing, but on the ability to input, store, process and output large quantities of data.

Later still, the electronic components that make up computers became sufficiently cheap for large quantities of relatively small, personal computers to be manufactured and sold, thus offering computing resources to a much larger section of the population. This also led to a boom in the demand for computer software and to the development of many new types of software, such as electronic spreadsheets, text processors, desktop publishers, small database programs, communications packages, graphical user interfaces etc. The interested amateur can now use such programs on his own computer hardware, along with traditional development software such as editors, compilers and linkers.

No longer is computing limited to purely numerical applications. For a majority of users, the floating-point performance of their hardware is of little importance.

Yet many still think that the applicability of the digital computer to mathematics merely consists in a speed-up of numerical computations. One of our aims is to prove that this is not the case. We shall try to convey how useful computer software, dedicated to *computer algebra*, the *symbolic* manipulation of mathematical *expressions*, can be to mathematicians, physicists, engineers and all others who use mathematical techniques from algebra and analysis.

Measured on the time scale of computer science, it is a very old idea to manipulate mathematical expressions—as opposed to approximate numerical quantities. The

1

very first computer algebra packages were written in order to automate particular sequences of mathematical operations on symbolic expressions.

The need for this comes from the wide application of mathematics to physics, chemistry, statistics and engineering. It has led, in the last two centuries, to the systematic formulation of 'laws' in mathematical terms, using algebraic equations or and analytic concepts such as ordinary and partial differential equations, integral theorems and variational principles. So whether you consider fluid dynamics, where

$$\frac{1}{2}\rho v^2 + p + \rho gh = C$$

or mechanics, where

$$\frac{d}{dt}\frac{\partial L}{\partial \dot{q}_i} - \frac{\partial L}{\partial q_i} = 0$$

or electromagnetism, where

$$\int_\Omega \vec{E} \cdot d\vec{S} = 4\pi q_\Omega$$

or quantum theory, where

$$(-\frac{\bar{h}^2}{2m}\vec{\nabla}^2 + U(x))\psi(x) = E\psi(x),$$

or any other quantitative law of physics or chemistry, it is expressed using algebra and analysis.

This use of mathematics has been essential for the advance of science and has yielded answers to many fundamental problems. Yet it is often quite a step from the formulation of a law (physics or chemistry) to its solution (mathematics).

For example, consider a point mass initially at rest in a homogeneous gravitational field. If x is its height, Newtonian mechanics leads to the ordinary differential equation

$$\frac{d^2 x}{dt^2} = -g,$$

where g is the constant acceleration caused by the gravitational field. The solution of this equation is easily seen to be

$$x = x_0 - \frac{1}{2}gt^2.$$

More precisely: it is easy because we have learned calculus. From this solution for x in terms of t, and from the principle of superposition of movement, one can deduce, for instance, the parabolic trajectory of a projectile, a fact that remained a mystery for centuries before the mathematical formulation of physics.

Now consider the same mass, but let it be placed in the gravitational field of a homogeneous spherical mass (such as the Earth, in good approximation). Then, x

denoting the distance between the point mass and the center of the spherical mass, the equation of motion becomes

$$\frac{d^2x}{dt^2} = -\frac{A}{x^2},$$

where A is a constant depending on the masses involved and on the choice of units. The solution of this equation can be found 'in closed form': we can write a true equality involving x, t, A and x_0, combined by a finite number of operations, and acted upon by functions selected from a finite set of 'elementary' ones (such as logarithms, exponentials, trigonometric functions). Explicitly,

$$t = \frac{1}{\sqrt{2A}}(x_0^{3/2}\arccos\sqrt{\frac{x}{x_0}} + \sqrt{xx_0(x_0 - x)}).$$

Note that this can no longer be solved in 'closed form' for x as a function of A, x_0 and t, unless one introduces ad hoc a new function.

The situation may be worse: very often, no closed-form solution of practical value exists. Sometimes, the solution can still be approximated using mathematical techniques. For instance, one may replace an unknown function by a power series and express the original equation as an infinite set of equations involving the series coefficients—hopefully in some usable, recurrent way. Or one can consider the equation at hand as a perturbation of an equation that can be satisfactorily solved, so that a parametrized family of equations can be introduced to study how the solution changes with the parameter, as it varies from the solved equation to the unsolved one. These techniques are widely used in applied mathematics, for example in quantum mechanics and in astronomy.

Computer algebra becomes necessary when the size of the expressions or the length of the computations grow beyond what a human computer can do in a reasonable amount of time and with a fair chance of not making mistakes.

The following anecdote illustrates this: the nineteenth-century French astronomer Delaunay spent twenty years working on a formula for the position of the Moon, computing the perturbation terms up to a sufficient order for the formula's accuracy to exceed the errors that are unavoidable when a physical measurement is made. Of these twenty years, he is said to have spent ten working out the formula, and the other ten checking his computations. Delaunay's formula was published and used successfully for several decades, nobody noticing that there were three erroneous terms in the result. These were only discovered when researchers solved the problem using computer algebra. Their specialized program required about twenty hours of computer time to find the single error Delaunay had made, which eventually had led to the different terms in the final result.

1.2 Requirements for a computer algebra system

Many researchers have implemented the basic functions required for computer algebra, on different computers, using different programming languages. There still are students and researchers who, at some point experiencing a need for computer algebra in their particular field, write programs for it. They might not invest the effort if they knew of the existence of computer algebra systems such as REDUCE, which not only address their needs, but can solve a wide class of mathematical problems, have been used intensively, and represent tens of man-years of development efforts.

We can only discourage what amounts to reinventing the wheel, but it is useful to examine how polynomials and rational functions can be manipulated using only a high-level programming language such as PASCAL. Advanced algorithms for integer and polynomial manipulations are discussed in detail in e.g. [1].

Let us start with polynomials. If only one variable (say, x) occurs in it, we could represent a polynomial

$$a_0 + a_1 x + \ldots + a_n x^n$$

internally by an array holding the values a_i. This leads to a nasty problem: there is no reason why the degree of the polynomial should be bounded a priori, but only arrays of fixed length can be declared in PASCAL. So it would be better to store the a_i in a *linked list*.

Now consider polynomials of a high degree, in which many terms vanish: it would be preferable not to store these terms. In that case, each element of the linked list must also hold the exponent of x.

The polynomial $3 - 2x + 16x^{45}$ could then be stored as

$$(3,0) \rightarrow (-2,1) \rightarrow (16,45)$$

taking up three items in the linked list, instead of forty-six. Note that the linked lists representing the polynomials should be ordered according to the exponent of x.

We can now sketch some algorithms for the symbolic manipulation of univariate polynomials.

- A first procedure would add a monomial to a polynomial.

  - It should verify that the monomial's coefficient is non-zero; otherwise nothing needs to be done.

  - Then it should scan the linked list to find an element with the same exponent of x.

  - If no such element can be found, the monomial should be copied into the linked list, which undergoes some surgery.

  - If an element with the same exponent is found, the monomial's coefficient should be added to the element's coefficient; if the resulting coefficient is zero, the element should be removed from the linked list and disposed of.

- Next, we write a function to return a copy of a polynomial.

  This amounts to copying the linked list.

- Now, we can write a function to add two polynomials.

  - It copies the first polynomial.
  - Then its scans the second polynomial and successively adds its monomials to the copy of the first one.
  - The modified copy is returned as the result.

- A function to compute (-1) times a given polynomial.

  It copies the polynomial and changes the sign of all coefficients.

- A function returning the difference of two polynomials.

  This uses $P - Q = P + (-Q)$.

- A function returning the product of two monomials.

  Allocate space for the result, multiply the coefficients, add the exponents.

- A function returning the product of two polynomials.

  - Start with the empty linked list, which represents the zero polynomial.
  - Scan the first polynomial and, for each of its monomials, scan the second polynomial.
  - For each pair of monomials, compute their product and add it to the result.
  - Return it.

- ...

Next, we must consider multivariate polynomials: the 'exponent' is no longer a single integer, but a linked list of variables and exponents. The polynomial manipulation functions must be modified accordingly. We might also choose a more efficient data structure to represent polynomials and exponents, a balanced tree for instance. We also must implement

- the substitution of a variable by a polynomial (which requires a function to compute the power of a polynomial)
- a way of taking partial derivatives of a polynomial
- univariate polynomial division
- a greatest common divisor algorithm for polynomials

- a factorization algorithm.

When all this would have been done, we could attack some more general mathematics. Recall our informal definition of 'closed form': inspection shows that any such expression can be written as a rational function (i.e. a quotient of multivariate polynomials), where the variables can be

- 'ordinary' ones such as x, y, A, t,

- or irrational constants such as π and e,

- or applications of irrational functions such as logarithms, exponentials and trigonometrics, on arguments that are themselves closed forms.

This is a recursive definition, so we require the nesting to be finite.

Example ——
This is a closed form expression of little mathematical utility, but some complexity:

$$\sqrt{2}\,\frac{15\sin(x^2 - y^2/7) + \pi\cos(x/y)}{x - y + 1} - \log(x + \exp(1/x)),$$

the 'variables' being x, y, π, $\sin(x^2 - y^2/7)$, $\cos(x/y)$, $\log(x + \exp(1/x))$ and $\sqrt{2}$, the last one because it is an application of the irrational function $\sqrt{\cdot}$ on the closed-form expression 2.

——

- Two such expressions can be added or subtracted using the rule

$$\frac{P_1}{Q_1} + \frac{P_2}{Q_2} = \frac{P_1 Q_2 + P_2 Q_1}{Q_1 Q_2}.$$

We could implement this at once using our polynomial manipulation functions.

- Multiplication is

$$\frac{P_1}{Q_1} \cdot \frac{P_2}{Q_2} = \frac{P_1 P_2}{Q_1 Q_2}.$$

- Invocations of irrational functions are kept:

$$\sin\frac{P}{Q} = \sin\frac{P}{Q}.$$

There is no general simplification of sin in rational terms.

- To differentiate a closed-form expression, we could use

$$\frac{\partial}{\partial \alpha}\left(\frac{P}{Q}\right) = \frac{Q P'_\alpha - P Q'_\alpha}{Q^2},$$

applying our polynomial differentiation function. When a 'compound variable' such as $\sin(x^2 - y^2/7)$ is reached, we could apply the chain rule, reducing the problem to the differentiation of the argument.

If we ensure that the set of elementary irrational functions S is such that whenever $f \in S$, f' is a closed form expression built with elements of S, we could differentiate any closed form expression, any number of times.

Here are some other problems we face:

- Our naïve approach to the addition of rational functions does not ensure that the greatest common divisor of numerator and denominator is always 1.

- We should include properties of the irrational functions, so that $\sin \pi = 0$ and $(\sqrt{2})^2 = 2$.

- The whole system should be user-extensible. One might, for instance, want to include the elliptic functions sn, cn, dn and their derivatives.

By then, we would have to decide at what stage each of the simplification rules should apply, when expressions should be factorized etc.
Another problem would be memory usage: our polynomial manipulation functions often create new elements in the linked lists, but very rarely dispose of them. After some computations, many of these elements will no longer be in use, yet remain allocated, unnecessarily taking up computer memory.

All that we described here can be done with REDUCE (and with several other packages). It indicates how complex even an elementary computer algebra package must be. REDUCE consists of about 50,000 lines of code, not taking into account the underlying LISP environment. Computer algebra systems can provide a functionality worth tens of man-years.

Over several decades, some of the pioneering computer algebra systems have evolved, through numerous improvements and several major rewritings, into state-of-the-art programs. This is especially true of REDUCE, which was originally written by A. Hearn to solve a problem in quantum theory. In its more than 20 years of existence, it has become one of the most widely used computer algebra systems, with an active user base still contributing to it.

1.3   Why REDUCE?

To be fair, we must mention several other interesting computer algebra packages, some of which are more recent than REDUCE:

- MACSYMA is a program that was developed at the Massachusetts Institute of Technology, as part of the MAC project. It is quite large, and written in LISP (just as REDUCE). Its size is a direct consequence of its completeness: it offers many features REDUCE does not have (yet), such as the solution of certain ordinary differential equations, limits, definite integration using complex analysis techniques, and tensor manipulation. Whereas REDUCE was originally designed by a single person, MACSYMA is the product of a group, and therefore many of its possibilities can be difficult to track down in its big manual. For instance, some functions are said to have a French name because they were first needed and implemented by a French researcher: do not look for int or ipart, it is called entier.

- MAPLE is a computer algebra package developed at Waterloo University (Canada). It is coded in C language instead of LISP, which makes it very fast, by avoiding some of the overhead created by LISP. Just as REDUCE, it runs on many different computer systems, including small ones. Its source code is not in the public domain.

- Some packages are 'computer algebra' in a somewhat different sense of the words: they provide computer support for algebraists. Of course, our 'closed form' expressions can be formulated in algebraic terms, but these packages tend to concentrate on mainstream algebra: group theory, modules, finite fields, Lie algebras, algebraic geometry etc. An important example is CAYLEY, an Australian package that was originally coded in FORTRAN and later automatically translated into C language. Other packages in this category, some of which are even more specialized, are MACAULAY (rings), LIE (Lie algebra), CoCoA (polynomial ideals) and GAP (groups).

- SCRATCHPAD-II is a package developed by IBM for its mainframes, UNIX workstations and PS/2 range of personal computers. It attempts to integrate computer algebra à la REDUCE and à la CAYLEY, along with new concepts in programming languages, using a strongly typed language with universal aspirations. It is implemented in LISP and not yet commercially available.

- MATHEMATICA is a package that offers advanced computer algebra along with excellent graphics on a variety of hardware platforms. It typically requires more computer resources than REDUCE, but less than MACSYMA. Its source code is not public.

- DERIVE and its precursor MuMATH are computer algebra packages that can run on small computers (8-bit, 64K RAM for MuMATH) and on computers whose memory addressing is awkward for LISP (such as IBM PCs based on the Intel 80x86 processors). Both are inexpensive and—within the limits of the hardware—very fast.

But we have chosen REDUCE. Indeed:

- REDUCE is widely available and widely used.

  In fact, REDUCE is built on top of a LISP dialect called 'Standard LISP', which was designed as a minimal standard for LISP. Under some form or other, it is available for almost all computers large enough to run it. This availability has been boosted further by the release of a Common LISP version of RE-DUCE: a simulation of Standard LISP in Common LISP brings a reasonable implementation of REDUCE to any computer having a Common LISP.

  There are several thousand official REDUCE sites.

- REDUCE is open.

  When you buy a copy of REDUCE, it includes the full source code for the IBM version. Often, this code is actually used to build the executable copy of REDUCE, so one can study and even modify it[1].

- REDUCE is relatively inexpensive.

  Most versions cost about US$500. That is rather expensive for a PC program, but very cheap for a workstation or mainframe program.

- REDUCE is relatively small.

  To do non-trivial work, 2MB of central memory is often sufficient (it depends on the LISP environment used). A personal computer capable of running RE-DUCE at a reasonable speed costs less than US$2000 (in 1991). Compare this to the 6MB of the MACSYMA dump file or the 4MB required by MATHE-MATICA: these programs can only run on workstations or expensive personal computers.

- REDUCE follows the trends in computer algebra, it is up to date.

  Because of its open character, developers like REDUCE. Their work is rapidly made available to the computer algebra community through electronic mail servers (currently, `reduce-netlib@rand.org` and `reduce-netlib@can.nl`).

---

[1]Such a modification is easily carried out because patching a LISP program often does not require a recompilation.

1.4   Classical high-level programming languages vs. REDUCE

Imagine that we have two computers at our disposal. We shall use them to illustrate some of the differences between LISP and computer algebra systems on the one hand, and high-level programming languages such as FORTRAN, PASCAL and C on the other.

We submit the following FORTRAN-77 program to the first computer:

```
PROGRAM test
REAL *4 X,Y,Z
READ(*,*) X,Y,Z
WRITE(*,*) X*(1./Y-1./Z)
END
```

After compilation and linking, we run it and supply the values
12.,2.,3.

The answer is
1.999999801

(note that user input will always be set in a slanted typeface, computer output in an unslanted one). The internal representation of REAL *4 is an approximation of real numbers. It is exact only for a finite set of rational numbers whose denominator is a power of 2. In particular, 1/3 cannot be represented exactly using this 'floating-point number' scheme, causing the error in the result.

How bad this is depends on the context. Floating-point numbers have proved their usefulness and the operations on them are much faster than their equivalents using true rational numbers (as in REDUCE). Furthermore, their memory requirements are fixed a priori, which is preferable in a static language such as FORTRAN-77. But they are only approximations, and the operations carried out on them do not have several of the properties which, mathematically, are fundamental for real numbers.

Example ───

The addition of floating-point numbers is not associative. If n is a sufficiently large integer, the floating-point sum of 1 and 2^{-n} will be 1. Denoting the floating-point addition by $\oplus$, we then have

$$(-1) \oplus (1 \oplus 2^{-n}) = (-1) \oplus 1 = 0$$

but

$$((-1) \oplus 1) \oplus 2^{-n} = 0 \oplus 2^{-n} = 2^{-n}.$$

───

When the quantities represented by floating-point numbers were obtained from necessarily imprecise measurements, one might think that these round-off errors are negligible, their relative size often being 2^{-31} or less in modern computers.

But this is not always true: when a numerical analyst devises an algorithm and implements it, he must take care to avoid the loss of important information by roundoff error during *all* stages of the program, taking into account that intermediate results might be much more affected by roundoff errors.

In some cases, the measurement errors which render the original data imprecise may also be correlated at the level of the mathematical representation, whereas round-off errors are not.

We start a REDUCE session on the other computer. At the prompt, we enter

12*(1/2-1/3);

and are answered

2

which suggests that REDUCE knows about fractions. To confirm this, we enter

2/3;

which yields

```
 2
---
 3
```

i.e. there is no simpler exact way of writing down 2/3.

Back to the first computer. We submit the following PASCAL program:

```
PROGRAM test(input,output);
VAR n: INTEGER;
FUNCTION factorial(n: INTEGER): INTEGER;
VAR i,result: INTEGER;
BEGIN {factorial}
        result:=1;
        FOR i:=2 TO N DO
                result:=result*i;
        factorial:=result
END; {factorial}
BEGIN {test}
        READLN(n);
        WRITELN(factorial(n))
END. {test}
```

Again compile, link and run. If we supply 3, the answer is 6. We run the program for a second time and enter 1000: the answer will be wrong, because the multiplication

```
        result:=result*i;
```

will repeatedly have overflowed. What is called INTEGER in PASCAL is only a limited range of numbers (typically, -2^n to $2^n - 1$ where n is 15 or 31).

Turn to the REDUCE session. The following is our first example of a REDUCE program—we shall describe the syntactic details in Chapter 3, but its meaning can readily be understood.

```
procedure fact n;
        for i:=2:n product i;
```

Notice how short this is, compared to the PASCAL program. Also note that RE-DUCE can be programmed interactively: no compilation or link step is required. Now we enter

```
fact 3;
```

the answer is

6

and if we enter

```
fact 1000;
```

the factorial of 1000 is computed and printed. It takes 2568 decimal digits and would fill a whole page, so we do not print it here. To represent it internally, at least 1100 bytes are needed—in a sense we have computed in INTEGER *1100 precision.

REDUCE has *true* integers[2], which are not limited by the machine word size. The only limitation on their absolute value is determined by the available memory. Here, the final result took between 1 and 2K of memory—imagine what enormous numbers might be stored in a multi-megabyte workstation! Seriously, the point is not the actual size of the numbers, but that the operations on them are precise and never overflow; in the unlikely event of integer arithmetic exhausting the computer memory, the problem would be signaled by REDUCE and the program would stop.

Back to the first computer. The following C language program reads characters until it reaches an A, then writes them out in reverse order:

```c
#include <stdio.h>
#include <malloc.h>
main()
{
        struct list {
                char c;
                struct list *previous}
        *last_element=NULL,*new_element,*previous_element;
        do {
                new_element=
                        (struct list *)malloc(sizeof(struct list));
```

---

[2]Often called *multi-precision integers* or *bignums* (for 'big numbers').

```
            if(new_element==NULL) {
                    fputs("Out of memory.\n",stderr);
                    exit(1);
            }
            new_element->c=getchar();
            new_element->previous=last_element;
            last_element=new_element;
    } while (last_element->c!='A');
    while(last_element!=NULL) {
            putchar(last_element->c);
            previous_element=last_element->previous;
            free(last_element);
            last_element=previous_element;
    }
    return(0);
}
```

Note how all memory allocation and deallocation is explicit, and how we must check, after each `malloc()` call, whether memory is not exhausted. We admit that the example is not realistic for a program of this size:

- many programmers would risk it not to check whether `malloc()` succeeded

- the deallocation at the end of the program is not necessary here

- many would dare to write

```
            free(last_element);
            last_element=last_element->previous;
```

irrespective of the fact that `*last_element` may have been modified by `free()`.

But in a robust C language program of non-trivial size, all these precautions must be taken, so the comparison between C and REDUCE would not be fair without them.

Again, a REDUCE equivalent. This is an RLISP program; the syntax will be described in Chapter 2. To clarify it, we point out that the `scalar` statement merely declares work as a temporary variable.

```
lisp procedure test;
begin
        scalar work;
        repeat
                work:=readch().work
        until eqcar(work,'a);
        for each v in work do
```

```
         princ v;
end;
```

It can be tested by typing

```
lisp test();
```

Again, the program is shorter and arguably easier to understand. The variable work holds a LISP list, the direct equivalent of the linked list pointed to by last_element in the C program. The function readch() reads the next character on the input stream; princ prints its argument, which is a character; eqcar(work,'a) is true if and only if the first element of the list work is A.

Now suppose you have a FORTRAN program in which the statement

```
     Z=3*X+4*Y
```

appears. Although it looks like an equation, it is an assignment: the compiler will generate code to compute 3*X+4*Y and store the result in Z.

In REDUCE, z=3*x+4*y behaves much more like an equation. For instance, it can be solved for z:

```
solve(z=3*x+4*y,z);
```

yielding the rather trivial

```
{Z=3*X+4*Y}
```

but also for x or y:

```
solve(z=3*x+4*y,x);
```

```
      - 4*Y + Z
{X=------------}
         3
```

It can be used to indicate that z should be replaced by 3*x+4*y in an expression, as in

```
sub(z=3*x+4*y,z**2);
```

```
     2                2
9*X   + 24*X*Y + 16*Y
```

and if we assign a definite value to y, say

```
y:=a+b;
```

the equation itself simplifies in terms of this value:

```
z=3*x+4*y;
```

```
Z=4*A + 4*B + 3*X
```

We see that REDUCE manipulates mathematical expressions symbolically, much as our imaginary package would have.

Let us try something more adventurous:

```
(x-1)*(x+1);
```

```
 2
X  - 1
```

This is the default behavior: when a product of polynomials is met, it is expanded and like terms are collected. REDUCE can also go the other way round, splitting a polynomial into irreducible factors. This non-trivial operation (remember high-school math exercises) can be useful whenever we need to determine the zeroes or the sign of an expression.

Example ————————————————————————————————

```
s10:=for i:=1:10 product x-i;
```

$$S10 := X^{10} - 55 \ast X^9 + 1320 \ast X^8 - 18150 \ast X^7 + 157773 \ast X^6 - 902055 \ast X^5$$

$$+ 3416930 \ast X^4 - 8409500 \ast X^3 + 12753576 \ast X^2 - 10628640 \ast X$$

$$+ 3628800$$

This is a Stirling polynomial—notoriously difficult to solve numerically, because a slight perturbation of the coefficients suffices to wipe out most real roots. But because we use exact arithmetic and because REDUCE contains several thousand lines of code for polynomial factorization, we can split it quite easily:

```
factorize s10;
{X - 10, X - 9, X - 8, X - 7, X - 6,
 X - 5, X - 4, X - 3, X - 2, X - 1}
```

(Do not think that REDUCE remembered the definition of s10 as a product! The factorization works just as well if s10 is entered manually.)

——

Finally, some lightweight analysis. REDUCE can take derivatives, such as

```
df(sin(x),x);
COS(X)
```

In fact, it can differentiate any closed form expression any number of times.
Integration is harder: there is no known easy and general way of finding a primitive, so we are taught a set of special methods: substitution, partial integration, what to do with trigonometrics etc. An important breakthrough in theoretical computer algebra has yielded a deterministic algorithm for integration, which is implemented in REDUCE.

Example ————————————————————————————————

```
int(1/(1-x**3),x);
```

$$2 \ast SQRT(3) \ast ATAN(\frac{2 \ast X + 1}{SQRT(3)}) + LOG(X^2 + X + 1) - 2 \ast LOG(X - 1)$$

———

To check this without retyping it, we use the symbol ws: it always denotes the last result

```
df(ws,x);
```

```
    - 1
  --------
    3
  X   - 1
```

In the following chapter, we shall present Standard LISP, the programming language in which REDUCE is written. Its relevance extends beyond REDUCE, or computer algebra as a whole: everyone can profit from some knowledge of LISP.

Chapter 2

Standard LISP AND RLISP

2.1 Computer programming

Computer memory is commonly organized as an array of byte-sized locations holding values that may be

- (parts of) program data

- (parts of) program instructions

- unused.

Physically, and at a sufficiently privileged level of the operating system, data and instructions are treated equally. The computer can process its own programs, load them, overwrite some of the instructions etc. This versatility is essential for the universality of digital computing. But it comes at a price: a lot can go wrong.

High-level languages have separated instructions and data conceptually, for reasons of safety, portability and expressivity. Yet one of the persistent errors occurring when programming in PASCAL or C language and using data structures of some complexity, is the so-called *wild pointer*: due to a programming error, a variable that was declared as a pointer to some other data element, actually holds an address

- that is not valid (a NULL for instance, or a wrongly-aligned address)

- or that points to an unused part of memory

- or to some non-initial part of a data element

- or to some data element of a different type than was declared.

A wild pointer may bring down the program when it is dereferenced, causing a crash or a dump, depending on the operating system. Such a program clearly is buggy, and the developer will hunt down the error. But some parts of a program may only rarely be used, and we have no general algorithm to design practical tests that detect *all*

17

potential errors in a program. A wild pointer that does not bring down the program
may lead to nearly untraceable data and instruction overwrites.

Luckily, some wild pointers can be detected by code criticizers such as lint. A classic
wild pointer occurs in

```
fprintf("%d",100);
```

where the output stream is omitted. Another source of wild pointers is the use of
non-initialized pointers of storage class auto, as in

```
a_function()
{
        char *p;
        *p=0;
        ...
}
```

Both bugs could be detected. But if we modify the last example to

```
a_function()
{
        char *p;
        do_nothing(&p);
        *p=0;
        ...
}
```

a code criticizer cannot know whether do_nothing modifies p: it might be part of
another code module, or of a binary library, and its initialization of p might depend
on run-time conditions.

These are C language examples; similar ones exist in PASCAL. FORTRAN has no
pointers, hence no wild ones, but this restriction limits the memory allocation to
a static maximum, which can be wasteful of memory resources and is awkward for
many applications ranging from systems programming to computer algebra.

LISP is often introduced as the language for *artificial intelligence*, as a realization
of *lambda calculus*, or as a tool for *functional programming*. Our justification for
it is far humbler: to us, LISP offers a safe way of using non-trivial data structures
interactively.

- It is safe because it prevents the creation of wild pointers (unless you are too
  clever and extend your LISP system with your own code[1]). The LISP environment cannot go down mysteriously.

- It is interactive because you are in a continuous dialogue with the LISP program.
  Compilation is optional and there is no linking step.

---

[1]With a BASIC-like poke function, say.

2.2   A model for LISP memory

We shall now describe how an imaginary LISP environment uses the computer memory at its disposal. Consider a large contiguous part of memory and visualize it not as an array of bytes, but as an array of larger entities called *dotted-pairs*. Dotted-pairs consist—as their name suggests—of two elements, both of which can independently be:

- a number (i.e. an integer[2] or a floating-point number[3])

- an identifier (written as a non-empty string *without* double-quotes).

  Examples: a, xyz1p30.

  Identifiers provide variables and constants in LISP. They can have a *value*, *properties* and *flags*. These terms will be clarified later on.

  Some identifiers are special: nil and t for instance. The value of these cannot be changed. nil has an important structural interpretation (Cf. ultra).

  The name of an identifier is automatically converted to upper case (actually, on some implementations, to lower case).

- a string (without value, properties or flags).

  Strings are written between double-quotes. If a double-quote occurs in the string, it is doubled, as in "An embedded double-quote "" is doubled". For portability, their length should not exceed 80 characters.

- a vector (the LISP equivalent of an array). In our imaginary LISP system these are not stored among the dotted-pairs, but in another contiguous part of memory. You can safely ignore them until we reach the LISP functions for vector manipulation. They are only rarely used.

- a function pointer (a reference to compiled LISP code). Usually, function pointers are not manipulated directly by the programmer. Their name is somewhat misleading, especially with respect to our remarks about wild pointers. Function pointers are always tame, and are technically speaking closer to *handles*, doubly-indirected pointers.

- a reference to another dotted-pair.

These data 'types' which can be elements of a dotted-pair are collectively known as *S-expressions*. For example, let the following be four consecutive dotted-pairs in memory:

---

[2]Integers are denoted in the format: [+-]?[0-9]+, i.e. an optional sign followed by a non-empty sequence of digits.

[3]Their format is [+-]?[0-9]*(([.][0-9]*)|[0-9])[E][+-]?[0-9]+. Examples: -123.456E78 and .123E-45.

Some of the notational conventions for dotted-pairs will be explained using this example.

- The first dotted-pair is conventionally written (a . b).

  It consists of the identifier a as first element and the identifier b as second element.

- The second dotted-pair has as its first element a reference to (a . b) and as its second element the number 15.

  Such a dotted-pair is conventionally written ((a . b) . 15), i.e. the reference to (a . b) is simply denoted by the dotted-pair being referenced.

- The last dotted-pair has as its first element the number 1 and as its second element the special identifier nil.

  It can therefore be written (1 . nil).

- The third dotted-pair, finally, can be written (v . (1 . nil)).

We shall explain shortly how the LISP user can create dotted-pairs, follow the references and represent data and programs.

The *list* is a LISP data structure which will be used to represent LISP programs and many LISP data elements. Any dotted-pair of the form

$$(S1 . (S2 . (S3 . (\cdots (Sk . nil)\cdots))))$$

where the Si are arbitrary S-expressions, is called a list. Also, the special identifier nil itself is defined to be a list, called the 'empty list'.

Because lists are used so often, there is a special notation for them: our prototypical list can also be written

$$(S1 \ S2 \ S3 \ \cdots \ Sk)$$

—no dots and far less brackets. The last two dotted-pairs in the figure are lists, and can be written (v 1) and (1).

Lists can be nested just as dotted-pairs can be. The same notational convention is used recursively. For instance, (a (b c) ((d))) is a list, which in dotted-pair notation would be written

$$(a . ((b . (c . nil)) . (((d . nil) . nil) . nil)))$$

A list is therefore a dotted-pair whose first element is the head of the list and whose second element is the rest of the list. To make this completely true, the special identifier nil is identified with the empty list and one may also write ( ) for nil.

2.3   Basic Standard LISP

Start a REDUCE session and type the command
end;
—this puts REDUCE in Standard LISP mode.

REDUCE is an interactive program: it waits for a command, executes it, prints the result to the screen and waits for the next command, ad infinitum (unless you stop it by typing 'bye;' or '(bye)', when in Standard LISP mode).

In LISP mode, if the input is an S-expression, Standard LISP will *evaluate* it and print the result. We shall gradually explain what this evaluation of S-expressions means.

- If we enter 15, the answer is simply 15 itself—numbers evaluate to themselves.

- Now type (quote a). The result is A.

  When a non-empty list is evaluated, and its first element is an identifier (here quote), the LISP system interprets this identifier as a function and performs it, taking the other elements of the list as its arguments.

  quote is a built-in LISP function, which takes a single argument and returns it at once, without any evaluation. This argument can be any S-expression. A notational convention allows you to write 'a for (quote a), '(1 2 3) for (quote (1 2 3)) etc.

Examples ————————————————————————

    (quote -9)

    -9

    (quote (a (b) (c d)))

    (A (B) (C D))

————————————————————————————————————

- The name 'evaluation' indicates that it allows one to go from a variable to its value. To exploit this, we must know how a variable can be given a value: using the built-in function setq (pronounced 'set-quote') .

Examples ————————————————————————

    (setq a 5)

    5

If we now enter

a

we get

5

---

setq takes two arguments, evaluates the second argument, assigns the result to the first argument, and returns it.

So if we next enter

(setq b a)

we get

5

and we can check that both a and b now have 5 as their value.

Note that setq does not evaluate its first argument:

(setq a 6)

will set the value of a to 6. If setq had evaluated its first argument, we would have attempted to set the value of 5 to 6, which would have produced an error message. (Try it: (setq 5 6).)

Using quote and setq, we can create S-expressions and assign them to identifiers. For example,

(setq a (quote (u v w)))

causes a to have (U V W) as its value.

Internally, the command has been parsed, a new list (U V W) has been created (using dotted-pairs), and has been assigned to a as its value. The required memory allocation (of at least three dotted-pairs) remains completely hidden.

LISP has many other built-in functions, to perform logical tests, arithmetic operations, structural operations, to change the value of identifiers etc. Furthermore, LISP allows you to add your own functions.
We shall:

- introduce a few arithmetic LISP functions in order to

- show how new LISP functions can be written by the user and

- finally, survey all built-in LISP functions.

The arithmetic operations in LISP are plus, difference, times, quotient and remainder.

When evaluated, (plus S1 S2 $\cdots$ Sk) causes the (arbitrary) S-expressions Si to be evaluated, and their resulting values to be added. Their sum is then returned. If an Si evaluates to a non-number (an identifier, a dotted-pair etc.), an error message is generated.

Examples ──
```
(plus 1 1)
2
(plus 10 -20 15)
5
(setq a 6)
6
(setq b 4)
4
(plus a 10 b)
20
(setq c 'a)
a
(plus 1 c)
***** Non-numeric argument in arithmetic
```
──

Similarly, when evaluated, (difference S1 S2) causes the S-expressions S1 and S2 to be evaluated; the resulting value of S2 is subtracted from the value of S1 and the result is returned.

If either value is non-numerical, an error message is produced.

Products can be evaluated just as sums, but using the LISP function times.

Examples ──
```
(times 2 3)
6
(times 1 2 3 4 5 6 7 8 9 10)
3628800
```
As with plus and difference,
```
(times 2 (quote a))
```
produces an error because (quote a) evaluates to a, which is not a number.
──

Finally, quotient and remainder are similar to difference in that they take two arguments, evaluate them, try to compute the quotient resp. the remainder and produce an error message if either argument does not evaluate to a numerical result. Furthermore, they fail if the second argument (which is the divider) evaluates to zero.

These five functions, along with setq and quote, are sufficient to write some small user-defined functions. There are different ways in which user-defined functions can be defined and used in LISP. Ordered by growing notational convenience:

- using an explicit lambda expression

- using an identifier which represents the lambda expression after a `putd` call

- using an identifier which represents it after a `de` call

- in RLISP, using an identifier which represents it after a `lisp` procedure definition.

The fundamental way uses *lambda expressions*. An example should clarify this method before we describe it in detail: suppose we want to compute the square of a number. If this number is the value of d, say (`setq d 6`) has just been entered, we can enter (`times d d`) to obtain 36. With a lambda expression, this can be done as follows:

`((lambda (x) (times x x)) d)`

—the result is again 36.

What has happened? Recall that in describing the evaluation of a list, we assumed that its first element was an identifier (such as `quote` or `plus`). Here, the list's first element is itself a list, viz (`lambda (x) (times x x)`). Such a list is called a lambda expression because:

- its first element is the identifier `lambda`

- its second element is a list of different identifiers which serve as argument prototypes

- the remaining elements of the list are S-expressions which will be evaluated consecutively, in such a way that the argument prototype identifiers have, as their value, the value of the corresponding arguments. The value of the last S-expression is returned.

Let us trace what happens when (`(lambda (x) (times x x)) d`) is evaluated.

- First, the arguments are evaluated from left to right. There is only one of them: d, and its value is 6.

- Then the evaluator sees that the first element of the list

  `((lambda (x) (times x x)) d)`

  is (`lambda (x) (times x x)`), a list instead of an identifier.

  This list starts with `lambda`, so it could be a lambda expression.

  Its second element is the list (`x`): it holds the single identifier x; therefore the lambda expression represents a function of a single argument.

- x is from now on understood to have the value 6.

- The evaluator computes (times x x) which is 6×6 or 36.

- There is no other S-expression at the end of the lambda expression, so 36 is the value returned.

- Just before this value is printed, the previous value of x is restored.

Lambda expressions seem to be a little complicated. Their essential advantage is that they are merely lists, S-expressions. There is no strict delineation between data and programs in LISP: both are S-expressions, kept in LISP memory using automatic memory allocation. This 'pure' use of lambda expressions is awkward. To compute another square, for instance, the lambda expression must be completely retyped.

A remedy is provided by the putd function ('put definition'). Using it, we can associate an identifier to a lambda expression in such a way that we shall be able to replace the lambda expression by the identifier.

This cannot be done using setq because the first element of a list is not evaluated: for instance,

(setq mysquare '(lambda (x) (times x x)))

(LAMBDA (X) (TIMES X X))

will *not* lead to

(mysquare 6)

yielding 36: mysquare is not evaluated when it occurs at the first location in a list. What should be done is:

(putd 'mysquare 'expr '(lambda (x) (times x x)))

MYSQUARE

Henceforth, (mysquare d) will return 36 if d has value 6, and we only have to type the lambda expression once, in the definition.

This definition of mysquare could be changed using another call of putd.

Notice the occurrence of the argument 'expr in the putd call. Standard LISP knows about three kinds of definitions that can be made using putd: expr functions, fexpr functions and macros. The last one will be discussed at a later stage. To distinguish between the first two types of functions, we introduce two conceptual properties a LISP function can have:

- We say that a LISP function is of 'eval' type if its result only depends on the value of its arguments.

  If it does not solely depend on their value (but perhaps on the identifiers occurring in them or on their form) it is called a 'noeval' function.

Examples ——

> plus, times, difference etc. are 'eval' functions—we mentioned that they
> add, multiply, subtract etc. the values of their arguments. On the other hand,
> quote cannot be an 'eval' function: even if a and b both have the same value
> (say, 6), (quote a) returns A and (quote b) returns B.

——

- We say that a LISP function is of 'spread' type if it takes a fixed number of
  arguments.

  If a function takes a variable number of arguments, these cannot be assigned
  to different argument prototypes, but are bundled in a list. Such a function is
  of type 'nospread'.

Examples ——

> plus and times cannot be 'spread' functions because they have a variable
> number of arguments. But mysquare, as defined above, is of 'spread' type. The
> same holds for hyposquare if we define it as follows:
>
> ```
> (putd 'hyposquare 'expr
>       '(lambda (x y) (plus (times x x) (times y y))))
> ```

——

Many built-in LISP functions are of 'spread' type.

The relation with expr and fexpr definitions is that:

- an expr definition yields a function of 'eval', 'spread' type

- a fexpr definition yields a function of 'noeval', 'nospread' type.

fexpr definitions are best avoided, because the support for them is not perfect when
using a Standard LISP which runs on Common LISP.

From what precedes, we see that very often putd will be used with 'expr and a
lambda expression. This warrants a shorter notation, which is offered by the de
function. Using de, our example function definition becomes

```
(de mysquare (x) (times x x))
```

i.e. de is followed by the identifier, followed by the tail of the lambda expression—
the lambda itself is omitted. This syntax for function definitions is almost universal
among LISP dialects, except that de might be replaced by another identifier (such
as defun).

Up to now, our user-defined functions only used their arguments prototypes (x in
mysquare, x and y in hyposquare). There is a way in LISP to use local variables,
too: using the prog function.

Example —————————————————————————————————

To compute $x + x^2$ given x, we can write
```
(de test (x) (plus x (times x x)))
```
but we can also assign the intermediate result x^2 to a temporary variable y, by writing:
```
(de test (x)
    (prog (y)
          (setq y (times x x))
          (return (plus x y))))
```

———

The syntax of the prog function is similar to that of a lambda expression: again there is a keyword (prog), followed by a list of identifiers—the names of the local variables—and by S-expressions.

There are a few differences, though: if one of these S-expressions is just an identifier, it is not evaluated but represents a *label* to which a jump can be carried out using the go function.

The value of a prog is determined when a return function is reached; if no such function is reached in the course of evaluation, the prog has value nil.

We have now covered enough introductory material to describe all built-in Standard LISP functions.

2.4   An overview of Standard LISP functions

(Cf. The Standard LISP Report [6]; this text is augmented with examples and permuted so as to separate Standard LISP specifics from more general LISP functions.)

2.4.1   Fundamental functions

The following functions are fundamental for any LISP system. We emphasize the importance of rplaca and rplacd and the details of structure sharing.

car

The function car takes a single argument. It is evaluated and should return a dotted-pair. The result of car is this dotted-pair's first element.

- Some LISP systems allow car also to apply on nil and define the value of (car nil) as nil.

  This is for instance the case in Common LISP. If used this way in a Common LISP version of REDUCE, no error will be signaled, but the resulting program may not run properly on other REDUCE systems.

  To the authors' opinion, this extension of car is not useful, because it consecrates the special role of nil at the core of the LISP system. Why could

one not choose another symbol for the empty list? nil is an identifier, not a
dotted-pair.

- The name car is not related to the vehicle but is shorthand for 'content of
address register', a meaning relevant to the first LISP implementations.

Examples ——

```
(car '(a b c))
A
(car '(a))
A
(car '(a . b))
A
(car '((1 2 3) a b c))
(1 2 3)
(car nil)
```
Try it out on your computer. . .
```
(car 'a)
```
This should give rise to an error message.

——

No copies are made by car. The fourth example can be illustrated as follows:

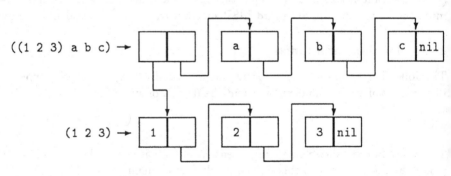

cdr

The function cdr takes a single argument, which is evaluated and should return a
dotted-pair. The result of cdr is the second element of this dotted pair.

- Some LISP systems allow cdr to act on nil, giving nil.

    This is the same situation as with car (Cf.). The same caution holds with
    respect to such usage.

- Some LISP systems allow cdr to act on any identifier and define its result as the property list (or property and flags list) of the identifier. This is not the case for Common LISP.

  Again, such use should be avoided if a portable REDUCE program is to be written.

  There is no clear reason why cdr should return the property list of an identifier. We surmise that this is merely a coincidence of the implementation, due to the fact that the property list of an identifier is stored (as a pointer) at an offset from the internal data structure representing the identifier equal the offset of the second element in a dotted-pair.

- The name cdr is pronounced 'could'er'. It means 'content of direct register'—a technical description relevant to the first LISP systems.

Examples ————————————————————————————————————

```
(cdr '(a b c))
```

(B C)

Bearing in mind the relation between lists and dotted-pairs, we see that cdr, acting on a non-empty list, returns its 'tail'.

```
(cdr '(a))
```

NIL

Again, this is because the last element of a list is represented by a dotted-pair whose second element is nil, or, equivalently, because the tail of (a) is the empty list nil.

```
(cdr '(a . b))
```

B

```
(cdr '((1 2 3) a b c))
```

(A B C)

```
(cdr nil)
```

Try it out on your computer. . .

```
(cdr 'a)
```

This may yield an error message, or a property list (which might be nil). In that case, to see a REDUCE object's property list, try (cdr 'sin) and enjoy deciphering it.

——

No copies are made by cdr. The first example can be illustrated as follows:

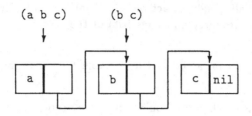

caar, cadr, cdar, cddr, caaar, caadr, cadar, caddr, cdaar, cdadr, cddar, cdddr, ... , cddddr

These functions each take one argument and evaluate it. The result is subjected to a series of car and/or cdr operations, depending on the function's name. Each of these car and cdr operations must apply to a dotted-pair.

Example ——

caddr is a function which takes one argument. This argument is evaluated. The result should be a dotted-pair whose second element is a dotted-pair, whose second element is a dotted pair; its first element is returned by the function. So (caddr <*expression*>) is equivalent to (car (cdr (cdr <*expression*>))), and a LISP implementor might use the definition (de caddr (x) (car (cdr (cdr x)))) (a direct encoding may be faster, though).

——

- The remarks on the possible extension of car and cdr to S-expressions that are not dotted-pairs, apply here too (Cf. car and cdr).

- Some of these function names can still be pronounced relatively easily (cadr for instance). Harder ones are cdar, cddr (could'ould'er), cdddr and cddddr. Also, cadar and cdar are difficult to distinguish in informal speech.

Examples ——

(cadr '(a b c))

B

We see that cadr is a way of finding the second element of a list which has at least two elements. Similarly, caddr returns the third element, and cadddr the fourth.

(cadr '(a))

This should return the same result as (car nil), because (cdr '(a)) has value NIL.

(cddr '(a b c))

(C)

```
(cdadar '((a (b c))))
(C)
```

No copies are made by these functions. The third example can be illustrated as follows:

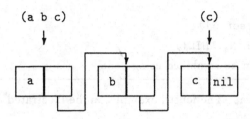

cons

The function cons takes two arguments. They are evaluated; the result of cons is a *new* dotted-pair with the value of the first argument as its first element and the value of the second argument as its second element. Any two S-expressions can be combined using cons.

- cons lets you build data structures in LISP, taking care of memory allocation. It is most often used to build lists. Because lists can be used to store sets, and because all structures are ultimately sets (relations also being sets), all data structures can, in some way, be built using LISP. But the data structures need not be lists. For instance, it is easy to build a binary tree using cons, each dotted-pair being a node.

- Some LISP systems offer, next to the 'ordinary' dotted-pairs built with cons, 'colored' ones called aconses, bconses etc. These can be analyzed using car and cdr just as ordinary conses, but instead of a single pairp (Cf.) predicate, there is a general consp predicate (true for conses, aconses etc.) and specific aconsp, bconsp etc. predicates for the colored ones. This feature is present in Cambridge LISP.

- The name cons is derived from 'construct'.

Examples ───

```
(cons 'a 'b)
(A . B)
(cons 'a '(b c))
(A B C)
```

This is the most frequent use of cons: it 'adds an element to the front of a list'.

(cons 'a nil)

(A)

This follows at once from the representation of lists as dotted-pairs.

(cons '(1 2) '(3 4))

((1 2) 3 4)

(cons "a string" (mkvect 3))

("a string" . [NIL NIL NIL NIL])

Any two S-expressions can be combined using cons.

No copies are made by cons. The fourth example can be illustrated as follows:

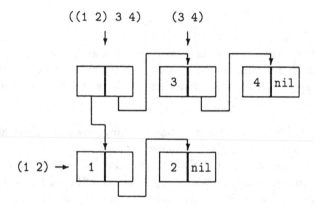

list

The function list takes any number of arguments (possibly zero) and evaluates
them. The results can be any S-expressions. The result of list is a new list which
has these results as elements, in the same order as they appeared in the list call.
If no arguments were supplied (i.e. the call was (list)), the result is the empty list
nil.

- The Standard LISP Report [6] offers the following definition of list:

  (df list (U) (evlis U))

  and therefore catalogues it as 'noeval', 'nospread'. We agree with 'nospread'
  (there is a variable number of arguments), but find 'noeval' unnecessary. To
  our opinion, list should be considered 'eval', 'nospread'.

Examples ───
```
(list 'a 'b 'c)
(A B C)
(list 'a)
(A)
(list nil)
(NIL)
(list)
NIL
(list '(1 2) '(3 4))
((1 2) (3 4))
```
───

No copies are made by list. The last example can be illustrated as follows:

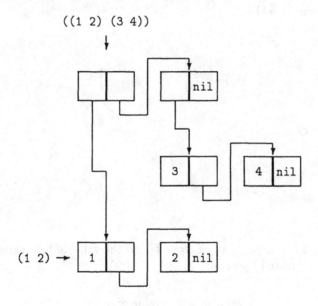

rplaca

The function rplaca takes two arguments. Both are evaluated. The result of the first one must be a dotted-pair. rplaca changes the first element of this dotted-pair to the result of the second argument. It returns the updated dotted-pair as its value.

- rplaca changes a particular dotted-pair. This is a very powerful function. Let, in a C language program,

```
struct pair {struct pair *first, *last;}
```

be a data type, and let pointer1, pointer2 be pointers to some elements in this structure. Then rplaca is similar to

```
((pointer1->first=pointer2),pointer1)
```

- The name rplaca is derived from 'replace car'.

- Because of the way copies are avoided by car, cdr, cons, list, append, member (Cf.) and other LISP functions, rplaca may affect more than was expected.

Examples ───

(setq a '(1 2))

(1 2)

(setq b (cons 0 a))

(0 1 2)

a

(1 2)

b

(0 1 2)

(rplaca a 'x)

(X 2)

a

(X 2)

b

(0 X 2)

Our rplaca on the value of a has also affected the value of b, because of the memory representation created by cons:

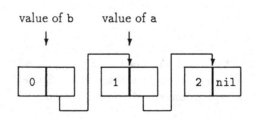

becomes

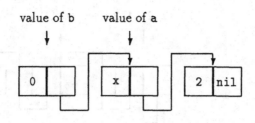

One might think this is a drawback of LISP. But the avoidance of copying is a great advantage for the performance (Cf. also our imaginary computer algebra package in Chapter 1). Also, it does not restrict the programmer's freedom in the way a universal copying scheme would. Finally, we often want this structure sharing (for instance when representing graphs).

rplaca can be used with cons to create certain special data structures.

Examples ────────────────────────────────

Make a dotted-pair whose two elements are the same list (1 2). A naïve way to try this is

(setq a '((1 2) . (1 2)))

((1 2) 1 2)

(car a)

(1 2)

(cdr a)

(1 2)

But now:
(rplaca (car a) 0)

(0 2)

we have
a

((0 2) 1 2)

We see that both elements of '((1 2) . (1 2)) are written (1 2), but that they are disjoint in memory. This can be illustrated as follows:

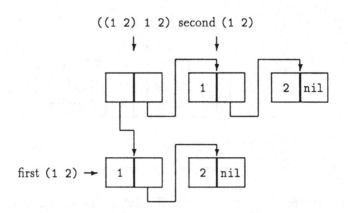

A correct solution runs as follows:

```
(setq b '(nil . (1 2)))
(NIL 1 2)
(rplaca b (cdr b))
((1 2) 1 2)
b
((1 2) 1 2)
```

Now we can check:

```
(rplaca (car b) 0)
(0 2)
b
((0 2) 0 2)
```

The memory representation for b is what we wanted:

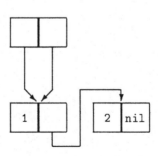

---

rplaca can also be used to make bizarre constructs, which may not print properly, or even hang the LISP interpreter.

Example ──────────────────────────────────────

Make a dotted-pair whose second element is 1 and whose first element is itself.
Solution:

```
(setq a '(1 1))
(1 1)
(rplaca a a)
```

You may get something strange, or an infinite sequence of opening parentheses, or nothing at all.
So this dotted-pair, whose memory representation is

cannot be printed. However, it is a perfectly valid LISP object, and can be used in computations, as long as they do not enter an infinite loop.
For instance,

```
(cdr ((lambda (x)
    (setq x '(1 1))
    (rplaca x x) x) nil))
1
```

where cdr acts (after the evaluation of the lambda expression call) on a structure that is similar to our a.
Using rplacd (Cf.) along with rplaca, one can make a dotted-pair whose first and second elements are itself:

```
((lambda (x)
    (setq x '(1 1))
    (rplaca x x)
    (rplacd x x) x)
 nil)
```

does exactly that.

──────────────────────────────────────

rplaca does not take copies.

rplacd

The function rplacd takes two arguments. They are both evaluated. The result of the first one should be a dotted-pair. rplacd changes the second element of this dotted-pair to the result of the second argument. This dotted-pair is the result of rplacd.

- rplacd changes an existing dotted-pair. This is a very powerful function. The parallel to it in C language is (with the definitions of the C code given in the description of rplaca)

$$((pointer1->second=pointer2),pointer1)$$

- The name rplacd is derived from 'replace cdr'.

- Again, the sharing of dotted-pairs in memory can lead to side-effects when rplacd is used.

Examples ————————————————————————————————————

```
(setq a '(1 2))
(1 2)
(setq b (cons 0 a))
(0 1 2)
(rplacd a '(u v))
(1 U V)
a
(1 U V)
b
(0 1 U V)
```

This can be illustrated as follows:

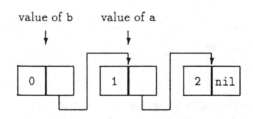

becomes

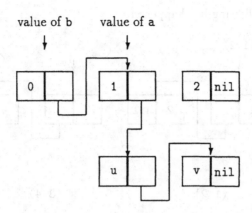

rplacd can be used, just as rplaca, to build bizarre dotted-pairs. For example, one can build seemingly infinite lists with it:

(setq a '(1 2))

(1 2)

(rplacd (cdr a) a)

This may yield something strange, or a list starting (1 2 1 2 1 2 and going on like that forever, or nothing at all. Again, we have created an object that is not safe to print, but can be used as a valid LISP object. This 'infinite list' has, in memory, the form:

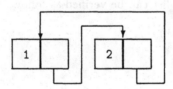

rplacd makes no copies.

append

The function append takes two arguments. Both are evaluated and should return lists. The result of append is a list holding the elements of the first argument's result, followed by those of the second argument's result. In the process, copies are made of the dotted-pairs that make up the first list, but not of those that make up the second.

Example ————————————————————————————————————

(append '(1 2) '(3 4))

(1 2 3 4)

In memory, the following has happened:

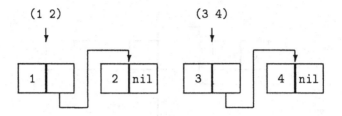

became

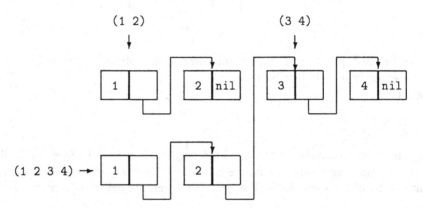

The sharing of the second list can be verified as follows:
(setq a '(1 2))
(1 2)
(setq b '(3 4))
(3 4)
(setq c (append a b))
(1 2 3 4)
(rplaca a 0)
(0 2)
a
(0 2)
c
(1 2 3 4)
—changing the first element of a's value has no effect on the value of c, because the
dotted-pair involved was copied by append. But
(rplaca b 0)
(0 4)

b

(0 4)

c

(1 2 0 4)

shows how the value of b is shared in memory with the cddr of the value of c.

- Some LISP systems allow append to take any number of variables. (append) might return nil, (append '(1 2 3)), (1 2 3) itself, and append of more than two lists may copy the dotted-pairs making up all the lists except the last. This is the case in Common LISP.

- Some other LISP systems do not produce error messages when append is used on more or less than two arguments, but have it return useless values. One system we encountered evaluates (append) as SIMPFN.

- Standard LISP does not provide a built-in function to copy structures. Therefore, (append <*list*> nil) is a convenient way to copy the dotted-pairs which make up <*list*>.

- The Standard LISP Report [6] offers the following recursive definition of append:

```
(de append (U V)
      (cond ((null U) V)
            (t (cons (car U)
                     (append (cdr U) V)))))
```

  It works even if the second argument does not evaluate to a list.

nconc

The function nconc takes two arguments. They are both evaluated and should return lists. nconc returns a list holding the elements of the first result, followed by those of the second result. In the process, the first list has been modified to become this result.

- nconc is functionally equivalent to rplacd of the last dotted-pair in the first list.

- the name nconc proceeds from an anonymous N and 'concatenate'.

- nconc is faster than append because no copies are taken. But it modifies the list obtained from its first argument.

- Some LISP systems allow nconc to take less or more than two arguments. The interpretations are the same as for append. This is the case for Common LISP.

- Some other LISP systems do not signal an error when nconc is used with a number of arguments different from two, but produce useless results.

Examples ───

```
(setq a '(1 2))
(1 2)
(setq b '(3 4))
(3 4)
(nconc a b)
(1 2 3 4)
a
(1 2 3 4)
b
(3 4)
```

In memory, the following has happened:

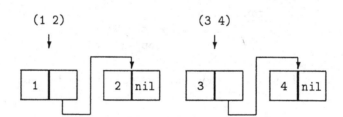

became

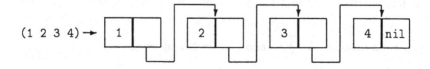

Because nconc does not copy the value of its first argument's value, it is more efficient to say (nconc '(1 2 3) x) than (append '(1 2 3) x) (we assume that x evaluates to a list), and both statements are completely equivalent when evaluated as such. Matters differ, however, in a procedure. Consider the function test:

(de test (x) (nconc (quote (1 2 3)) x))

When used for the first time, it works properly:

(test '(5))

(1 2 3 5)

But on the next occasion:

(test '(6))

(1 2 3 5 6)

which is not what we intended. This is caused by the fact that (1 2 3) in the definition of test is a fixed list that is modified by the nconc call. To solve this, we should ensure that a new list (1 2 3) is created each time it is required, for instance by the definition:

(de test (x) (nconc (list 1 2 3) x))

Then

test '(5);

(1 2 3 5)

test '(6);

(1 2 3 6)

But we could equally well have written

(de test (x) (append (quote (1 2 3)) x))

One can make an 'infinite' list by nconcing a list to itself, as in:

(setq a '(1 2))

(1 2)

(nconc a a)

This yields either something strange, a list starting (1 2 1 2 1 2 and going on forever, or nothing at all. Although the result of (nconc a a) in this example cannot be printed, it is a valid LISP object. Its memory representation is:

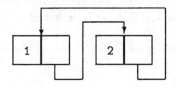

and we see that it is the same object as the one constructed when describing rplacd (Cf.).

reverse

The function **reverse** takes a single argument. It is evaluated and should yield a list. The value of **reverse** is a new list consisting of the elements of this list, but in reverse order.

Examples ————————————————————————————

(*reverse* '(1 2 3))

(3 2 1)

(*reverse* '((1 2) (3 4)))

((3 4) (1 2))

So **reverse** only reverses the list, not its elements that happen to be lists.
The second example, in memory:

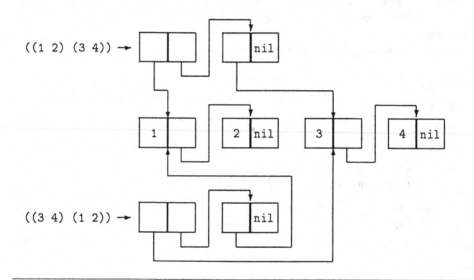

length

The function **length** takes a single argument. It is evaluated and should return a list. The number of elements in this list is returned.

Examples ————————————————————————————

(*length* '(1 2 3))

3

(*length* '(a))

1

(length nil)

0

If we construct an 'infinite list' using nconc or rplacd and call length on it, the LISP environment may hang or use up all memory and stop. Some LISP systems have clever alternatives to length which return nil in that case.

set

The function set takes two arguments. Both are evaluated and the first one should return a non-constant identifier. set sets the value of this identifier to the value of the second argument. The result of set is this value.
Some LISP systems allow set to take 2, 4, 6 etc. arguments and process them successively.

Examples ───

(set 'a 'b)

B

a

B

(set a 6)

6

a

B

b

6

(set 'c (cons a b))

(B . 6)

c

(B . 6)

---

setq

The function setq takes two arguments. The first one is not evaluated and should be a non-constant identifier. The second one is evaluated. The result is assigned as the value of the identifier and is also returned by the setq call.

- setq may be defined as a fexpr function:
        (df setq (u) (set (car u) (eval (cadr u))))

or as a macro:

$$(dm\ setq\ (u)\ (list\ 'set\ (list\ 'quote\ (cadr\ u))$$
$$(caddr\ U)))$$

- To use setq on a constant identifier should raise an error. Yet several LISP implementations do not, and may even crash if you change the value of nil or t. Van Hulzen [9] warns for this explicitly: never allow setq (or, for that matter, set) to change the value of nil or t.

2.4.2   Structural predicates

We have seen that S-expressions can be of different type: pairs, identifiers etc., and that some functions may only be used when their arguments are of a certain type. Often, we shall have to test whether a given S-expression is of a certain type. A function that performs such a test and either returns the value nil (for false), or a non-nil value (for true), is called a *predicate*.

The following diagram summarizes the built-in predicates of Standard LISP:

$$\left\{\begin{array}{l} \text{atom} \left\{\begin{array}{l} \text{constantp} \left\{\begin{array}{l} \text{numberp} \left\{\begin{array}{l} \text{fixp} \\ \text{floatp} \end{array}\right. \\ \text{vectorp} \\ \text{codep} \\ \text{stringp} \\ \text{null} \end{array}\right. \\ \text{idp} \end{array}\right. \\ \text{pairp} \end{array}\right.$$

Note that there is no predicate that tests whether its argument is a true list.

fixp

The function fixp takes a single argument. It is evaluated. If the result is an integer, t is returned, else nil.

- The name fixp alludes to a 'fixed number', i.e. an integer, as opposed to numbers that 'float'.

Examples ————————————————————————————————————
```
(fixp 1)
T
(fixp -62525241254)
T
```

```
(fixp 3.14159)
NIL
(fixp 'a)
NIL
(fixp '(1 2 3))
NIL
```

---

floatp

The function floatp takes a single argument. It is evaluated. If the result is a floating-point number, t is returned, else nil.

Examples ————————————————————————————————————

```
(floatp 1)
NIL
(floatp 3.14159)
T
(floatp 1.0)
T
(floatp 'a)
NIL
(floatp '(1 2 3))
NIL
```

---

numberp

The function numberp takes a single argument. It is evaluated. If the result is an integer or if it is a floating-point number, t is returned, else nil.

A definition of numberp in terms of fixp and floatp is
```
(de numberp (x) (or (floatp x) (fixp x)))
```

Examples ————————————————————————————————————

```
(numberp 1)
T
(numberp 3.14159)
T
```

```
(numberp 'a)
NIL
(numberp nil)
NIL
(numberp '(1 2 3))
NIL
```

---

vectorp

The function vectorp takes a single argument. It is evaluated. If the result is a vector, t is returned, else nil. (Vectors are described in section 2.4.8.)

Examples ───
```
(vectorp 1)
NIL
(vectorp 'a)
NIL
(vectorp '(1 2 3))
NIL
(vectorp (mkvect 6))
T
```
There is no way to enter a vector directly, so we had to use mkvect (Cf.).

---

codep

The function codep takes a single argument. It is evaluated. If the result is a code pointer, t is returned, else nil.

Examples ───
```
(codep 1)
NIL
(codep 'a)
NIL
(codep '(1 2 3))
NIL
(codep (function (lambda (x) (times x x))))
NIL
```

```
(codep (cdr (getd 'car)))
T
```
There is no way to enter a code pointer directly, so we have taken the function binding of car, assuming it is a compiled function.

---

stringp

The function stringp takes a single argument. It is evaluated. If the result is a string, t is returned, else nil.

Examples ————————————————————————————————
```
(stringp 1)
NIL
(stringp 'a)
NIL
(stringp '(1 2 3))
NIL
(stringp "a string")
T
```

---

null

The function null takes a single argument. It is evaluated. If the result is nil, t is returned, else nil.

Examples ————————————————————————————————
```
(null 1)
NIL
(null 0)
NIL
(null 'a)
NIL
(null '(1 2 3))
NIL
(null t)
NIL
(null nil)
T
```

---

idp

The function idp takes a single argument. It is evaluated. If the result is an identifier,
t is returned, else nil.

Examples ——————————————————————————————————
(idp 1)

NIL

(idp 0)

NIL

(idp 'a)

T

(idp '(a b c))

NIL

(idp nil)

T

(idp t)

T

———————————————————————————————————————

constantp

The function constantp takes a single argument. It is evaluated. If the result is a
'constant', t is returned, else nil.

By a 'constant', we mean an integer, a floating-point number, a code pointer, a
constant identifier, a string or a vector. For the latter two, this may need some
justification. In Standard LISP, strings cannot be modified, so it makes sense to
consider them as constants. But a vector can have its contents modified (using putv
(Cf.)). Vectors are treated as constants here because only the elements of the vector
can be modified, not the 'value' of the vector, which is always itself (i.e., eval of a
vector is that vector).

Examples ——————————————————————————————————
(constantp 1)

T

(constantp 0)

T

(constantp 3.14159)

T

```
(constantp (mkvect 6))
T
(constantp "a string")
T
(constantp (cdr (getd 'car)))
T
(constantp nil)
T
(constantp t)
T
(constantp 'a)
NIL
(constantp '(1 2 3))
NIL
```

---

atom

The function atom takes a single argument. It is evaluated. If the result is 'atomic', t is returned, else nil. An 'atomic' S-expression is defined as either a 'constant' one (Cf. constantp) or an identifier. Another way of defining atomic S-expressions is as follows: an S-expression is atomic if and only if it is not a dotted-pair. To consider vectors as 'atomic' is somewhat unfortunate.

Examples ———————————————————————————

```
(atom 1)
T
(atom 'a)
T
(atom nil)
T
(atom t)
T
(atom '(1 2 3))
NIL
```

---

pairp

The function `pairp` takes a single argument. It is evaluated. If the result is a dotted-pair, t is returned, else `nil`.

Examples ——
```
(pairp 1)
NIL
(pairp 1.345)
NIL
(pairp 'a)
NIL
(pairp nil)
NIL
(pairp '(1 . 2))
T
(pairp '(1 2))
T
```
——

2.4.3   Equality

Standard LISP offers three different equality tests: eq, eqn and equal (ordered by decreasing performance and increased laxity).

eq

The function eq takes two arguments. Both are evaluated. If both arguments yield the same object, t is returned, else `nil`. For eq, two objects are 'the same' if they occupy the same place in memory. Due to implementation details, two numbers with the same value may or may not be eq. An identifier is only eq with itself. A dotted-pair is also only eq with itself, not with a copy of itself. The same holds for vectors, strings etc.
eq can be implemented as a simple pointer test and can be compiled in-line. It may well be faster than eqn and is certainly faster than equal.

Examples ——
```
(eq 'a 'a)
T
(setq a '(1 2))
(1 2)
```

```
(eq a a)
T
(setq a "a string")
"a string"
(eq a a)
T
(setq a (mkvect 6))
[NIL NIL NIL NIL NIL NIL NIL]
(eq a a)
T
(eq '(1 2) '(1 2))
NIL
```
The two (1 2) are different copies in memory.
```
(eq (mkvect 6) (mkvect 6))
NIL
```
Ditto.

---

eqn

The function eqn takes two arguments. Both are evaluated. If the results are 'the
same object' in the sense of eq (Cf.), or if the results are both the same number
(integer or floating-point), t is returned, else nil. An integer cannot be eqn with a
floating-point number, not even if their mathematical values coincide.

Examples ———————————————————————————
```
(eqn 'a 'a)
T
(eqn '(1 2) '(1 2))
NIL
(eqn 1 2)
NIL
(eqn 1 1)
T
(eqn 3.14159 3.14159)
T
(eqn 1 1.0)
NIL
```

---

equal

The function equal takes two parameters. Both are evaluated.

- If one of the results is an identifier, an integer, a floating-point number or a code pointer, so must be the other and they must be eqn to return t, else nil is returned.

- If one of the results is a dotted-pair, so must be the other, their first elements must be equal and their second elements must be equal to return t, else nil is returned.

- If one of the results is a vector, so must be the other, they must have the same number of elements, and all these elements must be pairwise equal to return t, else nil is returned.

- If one of the results is a string, so must be the other, and their lengths and characters must be the same to return t, else nil is returned.

This definition does not fix the result of equal when applied to certain 'infinite' arguments. In these cases, equal may enter an infinite loop.

Examples ———

```
(equal '(1 2) '(1 2))
T
(equal (mkvect 6) (mkvect 6))
T
(equal "a string" "another")
NIL
(equal "a string" "a string")
T
```

2.4.4   Membership

member

The function member takes two arguments. Both are evaluated. The second one should return a list. member scans this list, trying to find a member that is equal to the first argument's value. As soon as such an element is found, the sublist starting with it is returned by member. If the search fails, nil is returned.

Examples ———

```
(setq l '(1 2 3 (1 2 3) 4))

(1 2 3 (1 2 3) 4)

(member 1 l)

(1 2 3 (1 2 3) 4)

(member '(1 2 3) l)

((1 2 3) 4)

(member 4 l)

(4)

(member 5 l)

NIL
```

——

memq

The function memq takes two arguments. Both are evaluated. The second one should return a list. memq scans this list, trying to find a member that is eq to the first argument's value. As soon as such an element is found, the sublist starting with it is returned by memq. If the search fails, nil is returned.

Examples ———

```
(setq l '(a b c (1 2 3) d))

(A B C (1 2 3) D)

(memq 'a l)

(A B C (1 2 3) D)

(memq '(1 2 3) l)

NIL
```

Both lists (1 2 3) are different in memory.

```
(memq 'd l)

(D)

(memq e l)

NIL
```

——

2.4.5   Functions related to identifiers

Recall (Cf. section 2.2) that to any identifier we can associate

- a value binding, established by set or setq

- a function binding, established by putd, de, df or dm

- flags

- properties.

Some LISP systems offer a function that returns the list of flags and properties associated with an identifier, or have some other means to access such a list (Cf. cdr). Standard LISP does not require such a function. We suppose this is in order to allow a faster implementation of the flags and properties (in a hash table, for instance). When there is a function that returns the actual property and flags list, an additional problem may occur when the user modifies it using rplaca, rplacd or nconc.

In general, an identifier cannot simultaneously have a value binding and a function binding.

Flags are represented by identifiers that name them; as their name indicates they represent a boolean value: either the flag is set or not. For example, we might set the flag beautiful for the identifier helen. Properties are more general than flags: a property of an identifier is named by an identifier and can have any S-expression as value. For instance, we could give helen the husbands property (menelaos paris), which is a list. In general, an identifier cannot at the same time have a flag and a property named by the same identifier.

flag

The function flag takes two arguments. Both are evaluated. The first should return a list of identifiers. The second should return an identifier. flag sets the flag given by the second argument's value for all elements of the first argument's value. The result of flag is the flag's name.

- It is a common mistake to forget that the first argument of flag must evaluate to a list, especially when only one identifier is to be flagged. So, (flag 'reduce 'useful) is an error; it should be (flag '(reduce) 'useful).

- Some LISP systems allow you to use non-identifiers such as integers as flag names. This must be avoided if portability is required.

Examples ───
```
(flag '(cat) 'furry)
```
FURRY
```
(flag '(haddock cod)
      'fish)
```
FISH
```
(flag nil 'something)
```
SOMETHING

You are allowed to flag the empty list of identifiers, i.e. do nothing.
```
(flag '(nil) 'empty)
```
empty

You are allowed to set flags on constant identifiers such as nil and t.
```
(flag '(bill) 'nil)
```
NIL

nil and t are valid as identifiers which name flags.
```
(flag '(marilyn mae marilyn)
      'actress)
```
ACTRESS

An identifier can appear twice in the value of the first argument. To set a flag once or twice or more is all the same (flags are not 'stacked', a flag twice set is removed by a single remflag call).

───

remflag

The function remflag takes two arguments. Both are evaluated. The first one should return a list of identifiers. The second one should return an identifier. remflag clears the flag named by the second argument's value, for all the elements of the first argument's value. The result of remflag is the name of the flag. It is not an error to clear a flag that was not set; the default settings of all flags is 'cleared'.

- The first argument to remflag must evaluate to a list of identifiers. Even if only one identifier must be affected by remflag, it must appear in a list, not by itself.

- If the LISP system allows flags that are named by non-identifiers (Cf. flag), these can also be cleared by remflag. Again, this is not portable on the level of Standard LISP.

Examples ───
The same as for flag, but with flag replaced by remflag.

───

flagp

The function flagp takes two arguments. Both are evaluated. If both results are identifiers, and if the flag named by the second is set for the first, t is returned. In all other cases, nil is returned.

- Either argument may evaluate to a non-identifier. The result of flagp is then nil, but no error occurs.

- If the LISP system allows non-identifiers as flag names, flagp also tests these, but this use is not portable on the level of Standard LISP.

Examples ──

(assuming that the examples given at the end of the discussion of flag have just been entered)

(flagp 'cat 'loud)

NIL

(flagp 'cat 'furry)

T

(flagp 'marilyn 'actress)

T

(flagp '(marilyn mae) 'actress)

NIL

The first argument did not evaluate to an identifier.

(flagp 'marilyn (mkvect 6))

NIL

The second argument did not evaluate to an identifier.

──

put

The function put takes three arguments. All three are evaluated. The first and second should return identifiers. put sets the property of the first, which is named by the second, to the third value, and returns this value.

- put cannot change the function binding of an identifier, because this is not a property but a separate data item.

- Some LISP systems may allow non-identifiers such as numbers to be used as property names. This is not portable at the Standard LISP level.

Examples ───

(put 'belgium 'languages '(dutch french))

(DUTCH FRENCH)

(put 'portugal 'languages '(portuguese))

(PORTUGUESE)

(put 'switzerland 'languages '(german french italian rhetoroman))

(GERMAN FRENCH ITALIAN RHETOROMAN)

Properties may also be named t or nil.

(put 'belgium 'languages '(dutch french german))

(DUTCH FRENCH GERMAN)

If a new value is given for a property, it simply overwrites the old one.

───

remprop

The function remprop takes two arguments. Both are evaluated and should return identifiers. remprop removes the property named by the second, from the first. It returns the removed property, or nil if no such property was present. It is not an error to remove a property that is not present.

- The Standard LISP Report [6] does not seem to require that both arguments evaluate to identifiers. But the authors' experience with several systems is that these do require it.

- If non-identifiers are allowed as property names, they can be removed using remprop.

Examples ───

(assuming that the examples at the end of the description of put have just been entered):

(remprop 'portugal 'languages)

(PORTUGUESE)

(remprop 'portugal 'languages)

NIL

───

get

The function get takes two arguments. Both are evaluated. If both return an identifier, and if the first has a property named by the second, this property is returned by get. Otherwise, nil is returned.

Examples ──

(assuming that the examples at the end of the description of put have just been entered):

(get 'belgium 'languages)

(DUTCH FRENCH GERMAN)

(get 'switzerland 'languages)

(GERMAN FRENCH ITALIAN RHETOROMAN)

Properties can be accessed with get any number of times, and may be very useful in programming because they make up a global database.

──

explode

The function explode takes a single argument. It is evaluated. The result should be an atom which is not a vector. The result of explode is a list of one-character identifiers coming from the printed representation of this atom. Some LISP systems allows explode to work on dotted-pairs and vectors too.

Examples ──

(explode 'LISP)

(L I S P)

(explode (times 7 9))

(!6 !3)

!6 is the identifier whose name is "6", not the integer value 6.

(explode "A STRING")

(!" A !   S T R I N G !")

Note that the double-quotes are still present.

(explode (cdr (getd 'car)))

See for yourself—it is a code pointer, whose printed representation is system-dependent.

──

compress

The compress function takes a single argument. It is evaluated and should return
a list of identifiers that are each one character long. The list is built into a LISP S-
expression, which will always be an atom and cannot be a vector. compress recognizes
numbers, strings and identifiers. Function pointers might be compressed, but this is
dangerous, since compress then can serve as a wild pointer factory. If compress does
not understand what you mean, it produces an error message.

It is a common error to write an integer (such as 3) in the list to be compressed,
instead of the corresponding digit (which would be !3). This error might not be
signaled, and bizarre results can follow.

Examples ——————————————————————————————————

```
(compress '(a b c))
```
ABC
```
(compress '(a !1 !2 b !3))
```
A12B3
```
(compress '(!" A !  S T R I N G !"))
```
"A STRING"
```
(compress '(!1 !2 !3 !4))
```
1234
```
(compress '(!1 !2 !. !3 !4))
```
12.3
```
(compress '(!1 !2 !. !3 E !- !5))
```
0.000123

———

gensym

The function gensym takes no arguments. It returns some identifier that has not yet
been used. gensym stands for 'generate symbol'.

Examples ——————————————————————————————————

```
(gensym)
```
G0001
```
(gensym)
```
G0002
etc.

———

gensym is often used in macros, for instance when a 'new' variable is needed as
a counter etc. Standard LISP ensures that the result of gensym is new by using its
'oblist', which holds all identifiers currently in use. REDUCE implementations that
simulate Standard LISP using some other LISP may warrant the following advice:
avoid the use of identifiers whose name is of the form [A-Z][1-9][0-9]* (some LISP
systems create symbols t1, t2 etc. instead of g1, g2 etc.)

intern and remob

Cf. the Standard LISP Report [6].

2.4.6   Function definition

putd

The function putd takes three arguments. All three are evaluated. The first should
return a non-constant identifier. The second should return the identifier expr, fexpr
or macro. The third should return a lambda expression or a code pointer. putd
assigns to the first argument a functional binding of type specified by the second
argument, and of value given by the third. If the global variable !*comp has a non-
nil value, putd will compile the third argument if it is a lambda expression, and
use the code pointer thus obtained. putd always returns the identifier naming the
function.

Examples ――――――――――――――――――――――――――――――――――――――
```
(putd 'myfun 'expr '(lambda (x) (times x x)))
MYFUN
(myfun 6)
36
(putd 'myexpr 'expr '(lambda (x) (prin2 x) '(plus 1 2)))
MYEXPR
```
(prin2 (Cf.) evaluates and prints its argument.)
```
(setq a 5)
5
(myexpr a)
5
(PLUS 1 2)
```
myexpr is an expr, i.e. an 'eval', 'spread' function:

- there is a single argument (supplying more or less arguments leads to an error),
  and

- only its value is used, so if both A and B evaluate to 5, (myexpr A) and (myexpr B) both evaluate to the same, viz the list (PLUS 1 2).

```
(putd 'myfexpr 'fexpr '(lambda (x) (prin2 x) '(plus 1 2))))
MYFEXPR
(myfexpr a)
(A)
(PLUS 1 2)
(myfexpr 5)
(5)
(PLUS 1 2)
(myfexpr '5)
((QUOTE 5))
(PLUS 1 2)
```

myfexpr is a fexpr, a function of type 'noeval', 'nospread':

- its arguments are not evaluated 'before they reach the lambda expression', so it is possible that (myfexpr A), (myfexpr 5) and (myfexpr '5) all behave differently when evaluated. Also,

- there is no fixed number of arguments: all arguments are bundled in a list and given to the lambda construct, which has only one argument prototype (here x).

```
(putd 'mymacro 'macro '(lambda (x) (prin2 x) '(plus 1 2)))
MYMACRO
(mymacro a)
(MYMACRO A)
3
(mymacro 5)
(MYMACRO 5)
3
(mymacro '5)
(MYMACRO (QUOTE 5))
3
```

mymacro is a macro. When evaluated,

- the lambda expression gets the arguments (which have not been evaluated) in a list, and preceded by the name of the macro itself. This is why (mymacro 5) leads to the list (mymacro 5) being printed.

- The lambda expression is evaluated. For mymacro, this always returns the list (plus 1 2).

- Finally, this result is once again evaluated (this second evaluation is characteristic of macros), and the final result is 3.

Because the lambda expression gets the whole of the macro call as its argument, macros are universal in the sense that, using their lambda expression, they can transform it in any way (introducing other macros etc.) before it is finally evaluated. This 'expansion' phase may take place during compilation, thereby allowing the user to apply complicated macros without worrying about their details. Another essential point is that these macros can use all the resources of LISP, whereas other macro schemes (such as the C preprocessor) use very restricted macro languages.

---

getd

The function getd takes a single argument. It is evaluated and should return an identifier. If this identifier has a function binding, getd returns a list whose first element is the type of function (expr, fexpr or macro) and the rest of which is either a lambda expression or a code pointer.

Examples ───

(getd 'pussycat)

NIL

There is no such function present yet.

(putd 'pussycat 'expr '(lambda nil 'meow))

PUSSYCAT

(getd 'pussycat)

(EXPR LAMBDA NIL (QUOTE MEOW))

(pussycat)

MEOW

(getd 'car)

Try it out for yourself. You should get a dotted-pair holding expr and a code pointer.

Remember that putd requires its second argument to evaluate either to expr, fexpr or macro. We cannot putd another type of function without incurring an error message. We can be devious and surgically modify the first element of the list obtained by getd:

(rplaca (getd 'pussycat) 'animal)

(ANIMAL LAMBDA NIL (QUOTE MEOW))

(getd 'pussycat)
(ANIMAL LAMBDA NIL (QUOTE MEOW))

But we can no longer use pussycat as a function:

(pussycat)

produces an error message stating that ANIMAL is not a valid function type.

remd

The function remd takes a single argument. It evaluates it, which should return an identifier. The function binding of this identifier is then removed; it is also returned by the remd call, in the same format as for getd. No error occurs if the identifier does not have a function binding; in that case, nil is returned.

de

The function de takes two arguments or more. None of them are evaluated. The first argument should be a non-constant identifier. The list formed by the second argument up to the last should, after addition of lambda in front, be a valid lambda construct. This means that the second argument to de must be a list of pairwise different non-constant identifiers. de gives its first argument the function binding (of expr) type, given by the lambda expression built from the remaining arguments by prepending a lambda.

Example ───────────────────────────────────────

(de myfun (x) (times x x))

defines an expr function myfun which evaluates and squares its argument. It is completely equivalent to

(putd 'myfun 'expr '(lambda (x) (times x x)))

───

The Standard LISP Report [6] suggests the following definition of de in terms of putd:

(putd 'de 'fexpr '(lambda (x)
        (putd (car x) 'expr (cons 'lambda (cdr x)))))

df

The function df takes two arguments or more. None of them are evaluated. The first argument should be a non-constant identifier. The list formed by the second argument up to the last should, after addition of lambda in front, be a valid lambda construct. This means that the second argument to df must be a list of pairwise different non-constant identifiers. df gives its first argument the function binding, of fexpr type, given by the lambda expression built from its other arguments by prepending a lambda to them. Because fexpr type functions are 'nospread', the lambda expression's list of argument prototypes should have exactly one element.

Example ──

```
(df myfexpr (x)
         (plus (times (car x) (car x))
               (times (cadr x) (cadr x))))
```

defines a fexpr function myfexpr that takes at least two arguments, does not evaluate them, and returns the sum of their squares. It is completely equivalent to

```
(putd 'myfexpr 'fexpr
    '(lambda (x) (plus (times (car x) (car x))
                       (times (cadr x) (cadr x)))))
```

An example of its use:

```
(myfexpr 3 4 anything)
25
```

──

The Standard LISP Report [6] suggests the following definition of df in terms of putd:

```
(putd 'df 'fexpr
    '(lambda (x) (putd (car x) 'fexpr (cons 'lambda (cdr x)))))
```

dm

The function dm takes two arguments or more. None of them are evaluated. The first argument should be a non-constant identifier. The list formed by the second argument up to the last should, after addition of lambda in front, be a valid lambda construct. This means that the second argument to dm must be a list of pairwise different non-constant identifiers. dm gives its first argument the function binding, of macro type, given by the lambda expression built from its remaining arguments by prepending a lambda to it.

Example ──

```
(dm mymacro (x) (apply (get (cadr x) 'mymacro) (list (cadr x))))
```

defines a macro function mymacro which takes at least one argument. It gets the mymacro property of this argument (which should be an identifier) and applies it to the identifier. So if we now enter

```
(put 'pussycat 'mymacro '(lambda (x) (prin2 x) '(plus 2 3)))
```

and

```
(put 'dog 'mymacro '(lambda (x) (list 'setq x 15)))
```

we have

```
(mymacro pussycat)
```

PUSSYCAT

5

and

(mymacro dog)

15

with, as a side-effect, the assignment of 15 to dog. The definition is completely equivalent to

```
(putd 'mymacro 'macro '(lambda (x)
        (apply (get (cadr x) 'mymacro) (list (cadr x)))))
```

The Standard LISP Report [6] suggests the following definition of dm in terms of putd:

```
(putd 'dm 'fexpr
        '(lambda (x) (putd (car x) 'macro (cons 'lambda (cdr x)))))
```

2.4.7   Logical functions

Boolean values are conventionally represented by nil for false and any non-nil value for true. If no useful value can be returned to signify 'true', the constant identifier t is used.

and

The function and takes any number of arguments. It evaluates them, starting by the first, until one of them returns nil. If this occurs, the following arguments are not evaluated and and returns nil. If all arguments return non-nil values, and returns the value of the last one (which cannot be nil and therefore correctly signifies 'true'). Called with no arguments, and returns t.

We see that and performs a 'lazy' logical conjunction. This laziness is the same as the one occurring in C language when the && operator is used. Yet and differs from && in that its value is not restricted to be 0 or 1, or nil or t, but is the value of the last of its arguments, when it is not nil. and differs more still from the and operator of PASCAL and from FORTRAN's .AND., which always evaluate both their arguments.

Examples ———————————————————————————————————

(and nil nil)

NIL

(and nil t)

NIL

(and t t)

T

```
(setq b 1)
1
(and (setq a nil) (setq b t))
NIL
b
1
```

Note that the (setq b t) has not been evaluated.

```
(and (setq a 1) (setq b 2))
2
a
1
b
2
```

---

or

The function or takes any number of arguments. It evaluates them, starting by the first, until one of them returns a non-nil value. If this happens, or returns this value. Otherwise, or returns nil. or called with no arguments returns nil.

We see that or performs a 'lazy' logical disjunction. This laziness is the same as that of the C operator ||. Yet or differs from || because its value is not restricted to be 0 or 1, or nil or t: when or is non-nil, it returns the value of the first of its arguments which evaluated to a non-nil result. or differs even more from the PASCAL or and from FORTRAN's .OR., because these always evaluate both their arguments.

Examples ————————————————————————————————

```
(or nil nil)
NIL
(or nil t)
T
(setq b 1)
1
(or (setq a t) (setq b 2))
T
b
1
```

Note that the (setq b 2) in the or call was not evaluated.
```
(or (setq a nil) (setq b 2))
```
2

b

2

---

not

The function not takes a single argument. It is evaluated. If the result is nil, not
returns t, otherwise not returns nil.
not is the same function as null. When used to perform a boolean operation, one
preferably uses not, and when performing a test whether some object is the empty
list, one preferably uses null.

Examples ————————————————————————————————

```
(not nil)
```
T
```
(not t)
```
NIL
```
(not '(1 t 6 2 d a))
```
NIL

---

cond

The function cond takes any number of arguments. Each argument should be a
non-empty list. Starting with the first argument, cond evaluates its first element. If
the result is nil, cond proceeds to the next argument, if any; if there is none, cond
returns nil. If the result is not nil, cond successively evaluates the other elements
of the list and returns the value of the last one. Called with no arguments, cond
returns nil.

Example ————————————————————————————————

```
(de test (x) (cond ((eq x 'a) "an a!") ((eq x 'b) "a b!")))
```
test
```
(test 'a)
```
"an a!"
```
(test 'b)
```
"a b!"

(test 'c)
NIL

So we see that cond performs conditional evaluation.

- It is a common mistake to forget parentheses when using cond.

- Sometimes, it is useful to have a catch-all case. This can be done by adding a final argument to cond, whose first element is t; this constant identifier always evaluates to a non-nil result.

  Our test example then becomes:
  (de test (x)
  (cond ((eq x 'a) "an a!")
  ((eq x 'b) "a b!")
  (t "what is this?")))

  and now
  (test 'c)

  "what is this?"

2.4.8 Vectors

Vectors are LISP data structures that can store references to S-expressions in a way more suited to direct access. If, for instance, you want to obtain the 500th element of a list, of which you only know the first dotted-pair, this will take 499 cdr calls and one car. There is no caddr⋯dddr function with 499 ds in it. And even if there were one, it would still take very long compared to cdr, because of the physical way in which LISP memory is arranged. To find the 500th element of a list will always take at least 499 pointer dereferences (unless some very strange LISP would be used).
Vectors answer this problem: to get the 500th element of a vector merely requires a multiplication of 499 by the vector cell size, and a reference is obtained. In LISP, vectors relate to lists as arrays relate to linked lists in PASCAL: to look up an element in a vector is often much faster than in a list, but lists can be extended at either end whereas vectors have fixed bounds.

mkvect

The function mkvect takes a single argument. It is evaluated and should return a non-negative integer. The result of mkvect is a vector with a number of locations equal to one plus the integer. All locations are initialized with nil. If the argument does not evaluate to a non-negative integer, an error is signaled. Making large arrays can strain the memory manager; there may be an implementation-dependent maximum array size, beyond which mkvect fails with an error message.

Examples ————————————————————————————
```
(mkvect 0)
[NIL]
(mkvect 6)
[NIL NIL NIL NIL NIL NIL NIL]
(setq a (mkvect 3))
[NIL NIL NIL NIL]
```
——

putv

The function putv takes three arguments. All three are evaluated. The first one
should yield a vector, the second one a non-negative integer which is a valid index
in the vector. The third can be any S-expression. putv changes the value of the cell
in the vector, corresponding to the integer (they are counted as 0, 1, 2 etc.), to the
S-expression. putv returns this S-expression.

Examples ————————————————————————————
```
(setq a (mkvect 3))
[NIL NIL NIL NIL]
(putv a 2 '(1 2 3))
(1 2 3)
a
[NIL NIL (1 2 3) NIL]
```
——

putv does not take copies.

getv

The function getv takes two arguments. Both are evaluated. The first should return
a vector, and the second a non-negative integer which is a valid index in the vector.
The result of getv is the element of the vector that corresponds to the integer.

Examples ————————————————————————————
(assuming that the examples at the end of the description of putv have just been
entered):
```
(getv a 0)
NIL
(getv a 2)
(1 2 3)
```
——

upbv

The function upbv takes a single argument. It is evaluated and should return a vector. upbv returns the number of elements in this vector, minus one.

Examples ————————————————————————————

(assuming that the examples given at the end of the description of putv have just been entered):
(upbv a)

3

(upbv (mkvect 16))

16

2.4.9   Program constructs

progn

The progn function takes any number of arguments. It evaluates them and returns the value of the last one. progn of no arguments returns nil.

Example —————————————————————————————

(progn (setq a 3) (setq b 5))

5

a

3

b

5

---

prog

The prog function takes one or more arguments.

- The first argument must be a list of pairwise different identifiers, which will be treated as local variables during the evaluation of the prog construct. If an argument of prog is an identifier, it is considered as a label, to which the program execution can jump using a go function; such a label is not evaluated.

- When the prog call is executed, the arguments are scanned, starting with the second, until an argument is found which is not a label. This argument is evaluated. If it is a call of the go function, prog considers the corresponding label (if it is not present, an error occurs), again looks for the first non-label argument to its right, evaluates it etc. If it is a call of the return function, the argument to return is evaluated and returned as the value of the prog call. Otherwise, the argument is evaluated and prog proceeds to the first argument following it, which is not a label, etc. If prog 'falls off' the list of arguments, the prog call returns nil.

Example ──

```
(prog (b)
      alabel
      (setq a 7)
      (go label3)
      label2
      (return (setq b 6))
      label3
      (go label2))
```

6

and, as a side-effect, the value of a has been set to 7. After evaluation of the prog, the value of b has not been set to 6, because b was a local variable in the prog, so (setq b 6) has no effect outside it.

──

go

The go function takes a single argument. It is not evaluated. It should be an identifier. go must be used inside a prog call, and the identifier must correspond to a label in this prog call, at the top level in the list (you cannot go from an inner prog to a label in an outer prog). go calls should only occur inside cond calls or progn calls, or at the top level. The detailed rules can be found in the Standard LISP Report [6].

return

The return function takes a single argument. It is evaluated. return must be used inside a prog call. It completes the prog call and has it return the value of its argument. return can only complete the current prog call, not another one at a higher level. The rules on the placement of return are the same as for go.

2.4.10 Numerical functions

plus2

The function plus2 takes two arguments. It evaluates them. Both should return a numerical value, i.e. either an integer or a floating-point number. plus2 adds them and returns the result. If either number is floating-point, the result will also be floating-point.

Examples ——————————————————————————————————————

(plus2 1 2)

3

(plus2 1.0 2)

3.0

(plus2 1.0 2.0)

3.0

———

plus

The function plus takes two or more arguments. It evaluates them successively. Each of them should return either an integer or a floating-point number. plus returns their sum. If any of them returned a floating-point number, the result will be a floating-point value.

Examples ——————————————————————————————————————

(plus 1 2 3 4)

10

(plus 1.0 2 3 4)

10.0

———

plus can be defined as a macro, as suggested in the Standard LISP Report [6]:
(dm plus (U) (expand (cdr U) 'plus2))

difference

The function difference takes two arguments. They are evaluated. Each should return either an integer or a floating-point number. difference computes their difference and returns it. If either number is floating-point, so is the result.

Examples ————————————————————————————————

```
(difference 2 1)
1
(difference 2.0 1)
1.0
```

---

minus

The function minus takes a single argument. It is evaluated and should return either
an integer or a floating-point number. minus returns (-1) times this number.

Examples ————————————————————————————————

```
(minus 1)
-1
(minus 1.0)
-1.0
```

---

times2

The function times2 takes two arguments. They are both evaluated. Each should
return an integer or a floating-point number. times2 multiplies them and returns
the result. If either number is floating-point, so is the result.

- times2 is not 'lazy' (Cf. and, or): it will evaluate the second argument even if
  the first returned 0 (integer zero).

- The result can be floating-point even if one of the factors was integer zero.

Examples ————————————————————————————————

```
(times2 2 3)
6
(times2 2.0 3)
6.0
(times2 0 (setq a 7))
0
```

and the value of A has been set to 7.

```
(times2 0 0.0)
0.0
```

---

times

The function times takes two or more arguments. It evaluates them. Each should return either an integer or a floating-point number. times returns their product. If one or more of the numbers were floating-point, so is the result.

Examples ───

```
(times 2 3 4)
24
(times 2.0 3 4)
24.0
(times 0 0.0 0 0.0)
0.0
```

───

The Standard LISP Report [6] suggests a macro definition of times:
```
(dm times (U) (expand (cdr U) 'times2))
```

divide

The function divide takes two arguments. Both are evaluated and should return an integer. The second should not be zero. divide returns a dotted-pair holding their quotient as its first element and their remainder as its second element.

- This description does not state exactly what is returned when one or both arguments evaluate to negative numbers. Such a specification can be found in the Standard LISP Report [6], but we point out that it is not followed in our implementation of REDUCE. You can only assume that the quotient and the remainder of positive numbers are the usual ones, and that the product of the quotient and the divisor equals the divided number minus the remainder.

- The Standard LISP Report [6] specifies implicitly that divide should work for floating-point numbers too. Some implementations of REDUCE refuse to compute the remainder of two floating-point numbers.

Example ──

```
(divide 7 4)
(1 . 3)
```

───

quotient

The function quotient takes two arguments. Both are evaluated and should return either an integer or a floating-point number. The second one should not evaluate to integer zero or floating-point zero. quotient returns the quotient of the numbers, which is an integer quotient if both are integers, otherwise it is a floating-point quotient.

The same caution applies as noted in the description of divide.

Examples ————————————————————————————

```
(quotient 1 2)
0
(quotient 1.0 2)
0.5
```

remainder

The function remainder takes two arguments. Both are evaluated and should return an integer. The second should not be zero. remainder returns their integer remainder. The same caution applies as noted during the description of divide.

- There is an inconsistency in the Standard LISP Report's definition of remainder in terms of quotient, because it would imply that remainder is floating-point zero when one or both arguments are floating-point, whereas the description of remainder promises something else, that would destroy the traditional relationship between divisor, divided number, quotient and remainder.

expt

The function expt takes two arguments. Both are evaluated and should return either an integer or a floating-point number. expt returns the first number raised to the power given by the second. If either number is floating-point, so is the result, but when an integer power of a floating-point number is computed, the integer is not converted to a float.

The Standard LISP Report [6] does not indicate the value of (expt 0 0).

Example ————————————————————————————

```
(expt 2 3)
8
(expt 2.0 3)
8.0
```

```
(expt 2 3.0)
8.0
(expt -2.0 3)
-8.0
```

---

abs

The function abs takes a single argument. It is evaluated and should return an integer or a floating-point number. abs computes the absolute value of this number and returns it.

Examples ————————————————————————————————

```
(abs 4)
4
(abs 4.0)
4.0
(abs -4.0)
4.0
```

---

max2

The function max2 takes two arguments. Both are evaluated and should return an integer or a floating-point number. If one of the argument values is strictly larger than the other, it is returned by max2, otherwise the first argument is returned.

Examples ————————————————————————————————

```
(max2 3 4)
4
(max2 3.0 4)
4
(max2 4.0 4)
4.0
```

---

max

The function max takes two or more arguments. All are evaluated; each should return an integer or a floating-point number. The largest value is determined and the first argument value equal to it is returned.

Examples ──

```
(max 7 3 2 8)
8
(max 2 5 2e10 20000000000)
2.0e+10
```

The Standard LISP Report [6] suggests the following macro definition of max:
```
(dm max (u) (expand (cdr u) 'max2)))
```

min2

The function min2 takes two arguments. Both are evaluated and should return either an integer or a floating-point number. If one value is strictly smaller than the other, it is returned as the value of the min2 call. Otherwise, the first value is returned.

Examples ──

```
(min2 1 2)
1
(min2 3 4.0)
3
(min2 4 4.0)
4
```

min

The function min takes at least two arguments. They are all evaluated. Each should give either an integer or a floating-point number. The smallest value is determined, and the first argument value which is equal to it, is returned.

Examples ──

```
(min 2 3 4)
2
(min 3 2.0 2 5)
2
```

lessp

The function lessp takes two arguments. Both are evaluated and should return an integer or a floating-point number. If the first value is strictly smaller than the second, t is returned, else nil.

Examples ————————————————————————————————

(lessp 1.0 3)

T

(lessp 1.0 1)

NIL

(lessp 1 1)

NIL

———

greaterp

The function greaterp takes two arguments. Both are evaluated and should return an integer or a floating-point number. If the first value is strictly larger than the second, t is returned, else nil.

Examples ————————————————————————————————

(greaterp 1 2)

NIL

(greaterp 3.0 1)

T

(greaterp 3.0 3)

NIL

———

fix

The function fix takes a single argument. It is evaluated and should return an integer or a floating-point number. If the value is an integer, it is returned by the fix call. Otherwise, the floating-point number is changed to an integer.

- The exact nature of this change for negative floating-point numbers is not specified by the Standard LISP Report [6],

- nor how very large floating-point numbers, which do not change when 1 is added to them, are handled.

`float`

The function `float` takes a single argument. It is evaluated and should return an integer or a floating-point number. If the value is floating-point, it is returned by the `float` call. Otherwise, the integer is transformed into floating-point format.

- On large integers, precision is lost.

- Very large integers cannot be represented as floating-point numbers.

2.4.11   Evaluation

`apply`

The function `apply` takes two arguments. Both are evaluated. The first should return an identifier which has a function binding, or a lambda expression, or a code pointer. The second should return a list. `apply` returns as its result the invocation of the function corresponding to its first argument value, when called on arguments which are listed in its second argument value.

- The type of the function binding (`expr`, `fexpr` or `macro`) is not considered here: the arguments are always delivered in a list.

Examples ───
```
(apply 'car '((1 . 2)))
1
(apply 'cdr '((1 . 2)))
2
(apply 'cons '(1 2))
(1 . 2)
```
───

`eval`

The function `eval` takes a single argument. It is evaluated. The resulting S-expression is again evaluated, and is the result of the `eval` call.

Example ──
```
(eval '(car (quote (1 2 3))))
1
```
`'(car (quote (1 2 3)))` evaluates to `(car (quote (1 2 3)))`, which evaluates to 1.
───

`evlis`

The function `evlis` takes a single argument. It is evaluated and should return a list. Each of the elements of this list is evaluated and the results are grouped in a new list, which is returned by `evlis`.

Example ───
```
(evlis '((times 2 2) (plus 3 4)))
(4 7)
```
───

`quote`

The function `quote` takes a single argument. It is not evaluated but returns immediately as the result of the `quote` call. (quote *<expression>*) may be abbreviated as '*<expression>*.
quote could be defined by
```
(df quote (x) (car x))
```
(but this would not raise an error when `quote` is called with more than one argument).

`expand`

The function `expand` takes two arguments. Both are evaluated. The first one should be a list. If *<expression 1>*, *<expression 2>*, ... , *<expression n>* are the elements of this list, and if *<value of expression 2>* is the value of the second argument, the result of `expand` is the list

$$(<value\ of\ expression\ 2>\ \ <expression\ 1>$$
$$(<value\ of\ expression\ 2>\ \ <expression\ 2>\ ...$$
$$(<value\ of\ expression\ 2>\ \ <expression\ (n-1)>\ \ <expression\ n>)...).$$

`expand` is useful in macros to define, given some associative operation `fn2` on two elements, its iterated counterpart `fn` on two or more. Cf. `plus` and `plus2`, for instance.

`function`

The function `function` takes a single argument. It is not evaluated, and should be either an identifier with a function binding, a lambda expression or a code pointer. Cf. The Standard LISP Report [6].
The purpose of `function` is to document the instances where LISP data (in particular, LISP lists having the lambda expression format) is considered as a function; it should assist the programmer in taking care of the necessary substitutions depending on the values of all identifiers at the moment when the `function` call is reached. To illustrate this, the following code should define a general composition function:

```
(de compose (f1 f2)
    (function
        (lambda (x)
            (apply f1 (list (apply f2 (list x)))))))
```

The function should ensure that, in the result, f1 and f2 are replaced by their values, since f1 and f2 lose their value binding when compose is terminated. Sadly, function does not generally live up to this in Standard LISP, so that, for instance,
(putd 'newfun 'expr (compose 'car 'car))

will not define newfun as an alternative to caar. This code would work in Common LISP, however.

2.4.12   Mapping functions

map

The function map takes two arguments. Both are evaluated. The first one should return a list, the second one a lambda expression, a code pointer or an identifier that has a function binding. map applies this function to the list, then to the cdr of the list, to its cddr etc. and stops just before nil. The result of map is always nil.

Example ──

```
(map '(1 2 3) '(lambda (x) (prin2 (times (car x) (car x)))))
1 4 9
NIL
```

───

mapc

The function mapc takes two arguments. Both are evaluated. The first one should return a list, the second one a lambda expression, a code pointer or an identifier that has a function binding. mapc applies this function to the car of the list, to its cadr, its caddr etc. The result of mapc is always nil.

Example ──

```
(mapc '(4 5 6) '(lambda (x) (prin2 (times x x))))
16 25 36
NIL
```

───

`mapcan`

The function `mapcan` takes two arguments. Both are evaluated. The first one should return a list, the second one a lambda expression, a code pointer or an identifier that has a function binding. `mapcan` applies the function to the `car` of the list, to its `cadr`, its `caddr` etc. and concatenates the results obtained from it. This is returned as the value of the `mapcan` call.

Example ───

```
(mapcan '(1 2 3) '(lambda (x) (list (times x x))))
(1 4 9)
```

───

Note that the concatenation is as in nconc, not as in append.

`mapcar`

The function `mapcar` takes two arguments. Both are evaluated. The first one should return a list, the second one a lambda expression, a code pointer or an identifier that has a function binding. `mapcar` applies the function to the `car` of the list, to its `cadr`, its `caddr` etc. and builds a new list holding the results, which is then returned as the result of the `mapcar` call.

Example ───

```
(mapcar '(4 5 6) '(lambda (x) (times x x)))
(16 25 36)
```

───

`mapcon`

The function `mapcon` takes two arguments. Both are evaluated. The first one should return a list, the second one a lambda expression, a code pointer or an identifier that has a function binding. `mapcon` applies the function to the list, to its `cdr`, its `cddr` etc. and concatenates the results obtained from it. This is returned as the value of the `mapcon` call.

Example ───

```
(mapcan '(1 2 3) '(lambda (x) (list (times (car x) (car x)))))
(1 4 9)
```

───

Note: the concatenation is as in nconc, not as in append.

maplist

The function maplist takes two arguments. Both are evaluated. The first one should return a list, the second one a lambda expression, a code pointer or an identifier that has a function binding. maplist applies the function to the list, to its cdr, its cddr etc. and builds a new list with the results obtained from it. This is returned as the value of the maplist call.

Example ───

```
(mapcan '(1 2 3) '(lambda (x) (list (times x x))))
(1 4 9)
```

───

Summary:

	return nil	build new list	concatenate
applied to the list, its cdr, cddr etc.	map	maplist	mapcon
applied to its car, cadr, caddr etc.	mapc	mapcar	mapcan

2.4.13   Association lists

An association list is a list whose elements are dotted-pairs each of which hold a key and a value.

assoc

The function assoc takes two arguments. Both are evaluated. The second should return an association list. assoc searches the association list until it finds an element whose first element is equal to the value of the first argument. If found, this element is returned, otherwise the value of the assoc call is nil.

Example ───

```
(setq l '((1 . 3) (in . no) ((2 3) . (5 6))))
((1 . 3) (IN . NO) ((2 3) 5 6))
(assoc 1 l)
(1 . 3)
(assoc 'in l)
(IN . NO)
(assoc '(2 3) l)
((2 3) 5 6)
(assoc 'how l)
NIL
```

───

sassoc

The function sassoc takes three arguments. All are evaluated. The second should
return an association list. The third should return a lambda expression, a function
pointer or an identifier with a function binding. If the first argument value can be
found as a key in the association list (in the sense of assoc), the corresponding
element of the association list is returned. Otherwise, the function is called without
arguments, and the result is returned by sassoc.

pair

The function pair takes two arguments. Both are evaluated and should return lists
of the same length. pair makes an association list in which the successive elements
of the first list occur as keys, with the corresponding elements of the second list as
values.

Examples ──

```
(pair nil nil)
NIL
(pair '(a) '(b))
((A . B))
(pair '(1 2 3) '(4 5 6))
((1 . 4) (2 . 5) (3 . 6))
```

───

2.4.14   Non-local jumps and error handling

Imagine you are writing a program in which the input is parsed by a function A. This
calls a function B for certain parts, which calls a function C, which calls a function
D. Now D discovers a syntax error that can only be handled by A. Then you might
have D return an error indicator as its value, to the function C which called it. This
function then immediately returns to the call in B, and B to A.

Such a scheme is necessary in languages like PASCAL. Its drawback is that each call
to a function must be followed by a test whether it did not return an error, and a
possible error return.

In C language, there is a simpler way of achieving the same, using a non-local goto
or 'long jump'. In function A, you save the current environment (i.e. the registers,
the program counter etc.) in a system-dependent structure called a jmp_buf. So, for
example:

```
void A()
{
```

```
        int the_error;
        if(the_error=setjmp(my_jmp_buffer)) {
                /* handle the error */
        } else {
                /* do what you want, for instance, call B */
        }
}

void B()
{
        /* what you want, call C */
}

void C()
{
        /* what you want, call D */
}

void D()
{
        /* some code */
        if( /* an error test */ ) {
                /* an error has occurred */
            longjmp(my_jump_buffer, /* an error code, say 15 */ );
        }
        /* code run when there was no error */
}
```

When A is entered, the setjmp call sets the contents of my_jump_buffer and, as a function, returns 0. This is conventionally interpreted as 'no error', and the if statement does not handle an error but executes the actions specific to A. This leads to B being called, which calls C, which calls D. In D, we check for a syntax error, say. If none occurs, D will return to C, C to B, and B to A. But if there is an error, we call longjmp (long jump) to restore the environment of A; the computer then behaves as if it had again returned from the setjmp at the beginning of A. But now, the value of setjmp is 15 (we gave this to longjmp), so we see that error 15 has occurred, and A handles it. All function calls below A can assume that no error occurs, because any error directly returns to the level of A.

This longjmp feature—which is an old assembly-language trick—can be very handy in many situations. Yet it gives some problems: assume we exit from A. The jump_buf might be a global variable; then we still could longjmp to it. But the particular environment of A (stack pointer etc.) is no longer valid, and this longjmp may cause

mysterious errors of the 'wild pointer' variety (code overwriting, data executed as if
it was code etc.)

LISP systems offer a similar, but safer construct. Often the functions involved are
called catch and throw. catch is the analogue of setjmp: it catches an S-expression
describing the error; throw is the analogue of longjmp: it throws the error description
towards catch. Just as, in C, several jump buffers can be valid simultaneously (one
might be set up by A, the other by B, so that when a syntax error is reached, D
decides whether A or B should handle it and longjmps to the appropriate jump
buffer), LISP systems allow catches and throws to be named and nested.

Standard LISP does not have this functions, but offers a similar, if slightly weaker
functionality using errorset and error. As their name indicates, they are mostly
used to report and handle errors, but their usage could be more general.

errorset

The function errorset takes three arguments. The first can be any expression.
errorset evaluates the first expression, but in such a way that if, during this evaluation, an error call is made, a non-local jump is made to errorset. Then, errorset
evaluates the second argument, and if it is not nil, displays the message sent by
the error call on the standard output and on the current output device (assuming
each is open). This message is displayed without top-level parentheses and preceded
by five asterisks. errorset then evaluates the third expression and, if it is not nil,
lists some trace-back information. The result of errorset is the error number of the
error call. If no error occurs during the evaluation of the first argument, its value is
returned by errorset.

error

The function error takes two arguments. Both are evaluated. The first one should
return an integer. The effect of error is to assign the second argument's value as the
value of the global variable emsg!* and to cause the corresponding errorset call to
return with the value of the first argument as its result.

2.4.15   Variable binding

There exist three types of variable binding in Standard LISP: global, fluid and local.

- The value of a global variable is the same regardless of where it is accessed;
  there is only one value for it. A variable must explicitly be declared global.
  Once a variable is successfully declared global, it can no longer be used as an
  argument prototype (in a lambda expression or, what amounts to the same,
  in a de, df or dm call), nor as a local variable in a prog construct, precisely
  because such use of this variable would imply that it has at least two possibly
  different values: one during the function's execution, and one outside it.

Global variables can be useful when programming, but most variables are fluid.

- Fluid variables are met most often. The default binding of a variable is fluid, so if the first use of a variable is not its declaration as a global variable, it will be fluid.

The value of a fluid variable depends on where it is evaluated. For instance:

Example ──

```
(setq a 15)

15

a

15

(de test1 (a) (prin2 a))

TEST1

(test1 67)

67
NIL

a

15
```

──

We see that A has value 15 'outside' test1, but that 'inside' it, the value of A is the value of the corresponding function argument—67 in the example. Once we have left the function, A recovers its previous value, 15.

The following illustrates this even better:

Example ──

```
(de test2 (a) (test3))

TEST2

(de test3 nil (prin2 a) nil)

TEST3

(test2 45)

45
NIL
```

Why is the value of A 45 in `test3`, and not 15? Because A appears as an argument prototype in `test2` this is the value of A during the call to `test2`, even for `test3`.

This also means that if we use a fluid variable instead of a global one, strange errors can occur just because two names are the same: that of a fluid variable, and that of an argument prototype used 'higher up'.

- Local binding occurs for the argument prototypes and prog local variables in compiled functions. Compilation speeds up program execution, partly because it can replace the fluid binding of variables by internal register allocation. So, in a compiled version of (lambda (x) (times x x)), x may well correspond to a register, and have no relationship with the identifier x; the argument prototypes and local variables in a compiled function are only placeholders and could be changed (within limitations: the name of a global variable cannot be used).

Example ──────────────────────────────────────

```
(setq !*comp t)
```

T

This tells putd to compile all future function definitions.
```
(de test2 (A) (test3))
```

TEST2
```
(de test3 nil (prin2 A) nil)
```

TEST3
```
(test2 45)
```

15
NIL

`fluid`

The function `fluid` takes a single argument. It is evaluated and should return a list of identifiers. These identifiers are declared fluid. Each identifier which had not yet been declared fluid is initialized to nil. A global variable cannot be changed into a fluid variable.

`fluidp`

The function `fluidp` takes a single argument. It is evaluated. If the result is an identifier and if it has been declared fluid using the function `fluid`, t is returned, otherwise nil.

global

The function global takes a single argument. It is evaluated and should return a list of identifiers. These identifiers are declared global. Each identifier which had not yet been declared global is initialized to nil. A fluid variable cannot be changed into a global variable.

globalp

The function globalp takes a single argument. It is evaluated. If the result is an identifier and if it has been declared global using the function global, t is returned, otherwise nil.

2.4.16   Other Standard LISP functions

deflist

The function deflist takes two arguments. Both are evaluated. The first should return a list whose elements are lists of two elements: an identifier and a general S-expression. The second argument should return an identifier. deflist sets the property named by the second argument's value, of each first element of an element of the first argument's value, to the second element of that element. deflist returns a new list holding the identifiers whose property was set.

Example ──

(deflist '((electron negative) (proton positive)) 'charge)
(ELECTRON PROTON)

This sets the charge property of electron to negative and the charge property of proton to positive.

──

delete

The function delete takes two arguments. Both are evaluated. The second one should return a list. delete scans the list until it finds an element that is equal to the first argument's value. delete then returns a list where this element is removed. The tail of this list (after the removed item) is shared with the original list, the head is new. If no equal member is found, delete returns the list itself.

Examples ───

(setq l '(1 2 3))
(1 2 3)

```
(setq m (delete 2 l))
(1 3)
(rplaca l 1)
(0 2 3)
l
(0 2 3)
m
(1 3)
(rplaca (cddr L) 4)
(4)
l
(0 2 4)
m
(1 4)
```

---

subst

The function subst takes three arguments. All of them are evaluated. subst performs a deep scan of the third argument and returns a copy of it where all (sub-)elements equal to the second argument's value are replaced by the first argument's value.

Example ───

```
(setq l '(1 (2 3) 2 (3 (2 3))))
(1 (2 3) 2 (3 (2 3)))
(subst '(5 6) '(2 3) l)
(1 (5 6) 2 (3 (5 6)))
```

The copy of the third argument's value is taken in the naïve way, so that shared structure is not preserved.
Let us verify this:

```
(setq w
    ((lambda (x)
        (cons x x))
      '(6)))
((6) 6)
```

w has internally shared structure: its cdr and car are eq, so changing one of them affects the other:

```
(rplaca (cdr w) 1)
(1)
```

```
w
((1) 1)
```
Now we call subst:
```
(setq w1 (subst 5 1 w))
((5) 5)
```
But the naïve copy has no internally shared structure:
```
(rplaca (cdr w1) 6)
(6)
w1
((5) 6)
```

---

sublis

The function sublis takes two arguments. Both are evaluated. The first one should return an association list. sublis performs a deep scan of the second argument's value and builds a copy of the third argument's value, but whenever it encounters a (sub-)element which is equal to some key in the association list, it replaces it in the new structure. This is finally returned.

Example ———
```
(sublis '(((1 2 3) . (4 5 6)) ((a b c) . (d e f)))
        '((1 2 3) (a b c) ((1 2 3) (a b c))))
((4 5 6) (D E F) ((4 5 6) (D E F)))
```

The copy of the third argument's value is naïve in the sense noted during the description of subst.

liter

The function liter takes a single argument. It is evaluated. If it returns an identifier whose name is a single upper or lower-case letter, t is returned, otherwise nil.

Examples ———
```
(liter 'a)
T
(liter 'aa)
NIL
(liter 1)
NIL
```

---

digit

The function digit takes a single argument. It is evaluated. If the result is an identifier whose name consists of a single character that is a digit, t is returned, otherwise nil.

Examples ————————————————————————————————
(digit 'a)
NIL
(digit 1)
NIL
(digit '!1)
T

———

close

The function close takes a single argument. It is evaluated and should return a file handle as obtained from an open call. The corresponding file is properly closed. close returns the value of its argument.

open

The function open takes two arguments. Both are evaluated. The second one should return either input or output. The first value must be an implementation-dependent name for the file that should be opened (this can be an identifier or a string, for instance). open attempts to open the file for input or output, depending on the second argument's value. Open returns an S-expression that can be used as an argument for rds and wrs.

rds

The function rds takes a single argument and evaluates it. The result must have been obtained from an open call or be nil. rds suspends the input from the current stream and switches to the stream associated with its argument: either the file that was opened (and not closed yet), or standard input if the argument's value is nil. rds returns the internal representation of the previous input stream.

wrs

The function wrs takes a single argument and evaluates it. The result must have been obtained from an open call or be nil. wrs suspends the output to the current stream and switches to the stream associated with its argument: either the file that was opened (and not closed yet), or standard output if the argument's value is nil. wrs returns the internal representation of the previous output stream.

princ

The function princ takes a single argument. It is evaluated and should return an identifier whose name consists of a single character. This single character is printed by princ. The value is the identifier. Remark: do not (princ 1) etc. but (princ '!1) etc.

Example ———————————————————————————————
```
(princ 'a)
A
A
```
——————————————————————————————————————

prin1

The function prin1 takes a single argument. It is evaluated. prin1 prints it in such a way that read could read it back without any change (except for internal shared structure). The argument's value is returned by prin1.

- read cannot always read back the output of prin1: it will fail for vectors and possibly for code pointers.

Examples ———————————————————————————————
```
(prin1 'a)
A
A
(prin1 015624)
15624
15624
(prin1 !015624)
!015624
!015624
(prin1 "a string")
"a string"
"a string"
(prin1 '(1 2 3))
(1 2 3)
(1 2 3)
```
——————————————————————————————————————

prin2

The function prin2 takes a single argument. It is evaluated. The result is printed in
a way that generally cannot be read back by read. Strings are not double-quoted by
prin2. Special characters in identifiers are not escaped by exclamation marks. The
argument's value is returned by prin2.

Examples ————————————————————————————————————

(prin2 'a)

A

A

(prin2 '!0123)

0123

!0123

(prin2 "a string")

a string

"a string"

——

print

The function print takes a single argument. It is evaluated. print prints the result
exactly as prin1 would, but adds a new-line at its end.

terpri

The function terpri takes no arguments. It returns nil. When it is used, a new
output line is taken.

• terpri stands for 'terminate printing' (of the current line).

readch

The function readch takes no arguments. It returns the next character from the
current input file, as an identifier with a one-character name. At end-of-file, it returns
the value of !$eof!$.

read

The function read takes no arguments. It reads the next expression from the current
input file. At end-of-file, it returns the value of !$eof!$.

pagelength

The function pagelength takes a single argument. It evaluates the argument, which should return an integer. It sets the vertical length (in lines) of the output page to this integer, in order to eject automatically each full page. The integer is returned.

posn

The function posn takes no arguments. It returns the number of characters present in the output buffer.

eject

The function eject takes no arguments and returns nil. It causes the current output page to be ejected.

lposn

The function lposn takes no arguments. It returns the number of lines printed on the current output page.

2.5   RLISP and Standard LISP.

RLISP is the 'first' part of REDUCE, to be found in the rlisp.red source file. It implements a number of support routines for REDUCE (which we only mention) and the possibility to use a prefix syntax close to Algol instead of the parentheses-LISP in general use. The main instrument for this is the function xread, which is the RLISP equivalent of read in Standard LISP.
A formal characterization of the transformations carried out by xread can be found at the end of the Standard LISP Report [6] and, of course, in the source code itself. We only illustrate the main advantages of RLISP over Standard LISP. Bear in mind that these differences are only syntactic; the actual power of both is the same.
RLISP is partly designed to avoid brackets. Function calls can be written in prefix form, so
myfun(a,b)
is the RLISP equivalent of (myfun a b) in Standard LISP.
If there is only one argument, the parentheses can be dropped, so
car x
is the RLISP equivalent of (car x). Beware that function application has a very high precedence, so in RLISP
car x + 1
means car(x)+1, or (plus2 (car x) 1) in Standard LISP.

RLISP also has parentheses-free, infix versions for the basic operations:

`a + b + c`

in RLISP corresponds to (plus a b c) in Standard LISP. The same for `times`, `quotient`, `minus`, `difference` and `expt`. `cons` is also a basic operation, with infix form '`.`':

`a . b`

corresponds to (cons a b). RLISP also knows <, >, =, <= and >=, which correspond to `lessp` etc. In particular, = corresponds to `equal`, not to `eq` or `eqn`, but you can use `eq` and `eqn` as infix operators. The LISP functions `and` and `or` can also be used as infix operators. You are not forced to use these infix operators: for instance, `plus(a,b,c)` is just as valid in RLISP as a+b+c. Assignment can also be done in prefix form: `a:=56` corresponds to (setq a 56). There is also a way to perform `rplaca` and `rplacd` in infix form, illustrated as follows: `car a:=67` corresponds to (rplaca A 67). Note that a is evaluated in this case.

`cond` is replaced by the RLISP `if` statement.

Example ───

`if a<6 then x:=78;`

corresponds to

`(cond ((lessp a 6) (setq x 78)))`

and

`if a<6 then x:=78 else y:=65;`

corresponds to

`(cond ((lessp a 6) (setq x 78)) (t (setq y 65)))`

───

RLISP also gets rid of the parentheses produced by progn and prog. To join statements into a progn, simply write them one after another, separated by semicolons, and enclose them between << and >>.

Example ───

`<<x:=y-6; z:=x*x>>`

corresponds to

`(progn (setq x (difference y 6)) (setq z (times x x)))`

in Standard LISP. The value of this progn is that of the last statement. Note that this is not followed by a semicolon. If one or more semicolons are present, as in

`<<x:=y-6; x:=x*x;>>`

the corresponding Standard LISP is

`(progn (setq x (difference y 6)) (setq z (times x x)) nil)`

and we see the nil at the end of the progn. This construct always has value nil.

───

To put statements into a prog form, write them between a begin ... end pair. Local variables can be obtained by a scalar statement, which must immediately follow the begin. Labels are followed by a colon. go statements are in the form go <*label*> or goto <*label*>. return statements are in the form return <*expression*>.

Example ———————————————————————————————————————

```
begin
    scalar x,y;
    x:=u-v;
    if x>0 then return u
            else goto 12;
l1:
    return v;
l2:        x:=x+1;
end;
```

corresponds to

```
(prog (x y)
      (setq x (difference u v))
      (cond ((greaterp x 0) (return u))
            (t (go 12)))
l1
      (return v)
l2
      (setq x (plus x 1)))
```

The value of a begin ... end block is therefore that of the return statement which terminated it, or else nil.

Lambda expressions can still be entered as '(lambda ...) in RLISP, but may also be written in the form
lambda <arguments>; <statement>;
Note that only one statement may occur here, but if several are needed they are easily bundled using << ... >> or begin ... end.

Example ———————————————————————————————————————

```
(lambda (x); x*x) 6;
```

corresponds to

```
((lambda (x) (times x x)) 6)
```

Here, lambda(x) can also be written lambda x.

RLISP replaces de, df and dm by the syntax

```
[ expr | fexpr | macro | ] procedure <identifier> (<arg 1>,...) ;
                        <procedure body>
```

Example ——

The RLISP definition

procedure myfun(x,y); x+y+x*y;

corresponds to the Standard LISP

(de myfun (x) (plus x y (times x y)))

and the RLISP definition

macro procedure mymac(x); get(car x,'myprop);

corresponds to

(dm mymac (x) (get (car x) 'myprop))

in Standard LISP. Here too, mymac(x) could be written mymac x.

——

RLISP also has while and repeat statements, with syntax

$$\text{while } <expression> \text{ do } <statement>$$

and

$$\text{repeat } <statement> \text{ until } <expression>$$

Both are transformed into corresponding prog constructs.
RLISP has different kinds of loops:

$$\text{for } <identifier> := <expression\ 1> \text{ to } <expression\ 2> \text{ step } <expression\ 3>$$
$$\text{do } <statement>$$

$$\text{for } <identifier> := <expression\ 1>:<expression\ 2> \text{ do } <statement>$$

$$\text{for each } <identifier> [\text{ in } | \text{ on }] <expression> [\text{ do } | \text{ collect } | \text{ join }]$$
$$<statement>$$

The first two perform numerical loops. The identifier becomes a local variable; these
statements are changed into progs. The last one performs an iteration over a list.
If in is chosen, the identifier is given the value of the car of the expression, then
of its cadr, caddr etc. If on is chosen, these values are the list itself, its cdr, cddr
etc. If do is chosen, the statement's value is nil. If collect is chosen, it consists
of a new list built from the values of *<statement>*. If join is chosen, these results
are concatenated. This for each statement is transformed into a mapping function
(mapcar for in ... collect etc.).

There is a general method to implement such RLISP statement forms. If the RLISP
interpreter sees a command starting by an identifier which has a non-nil stat property, it gets this property (which should be a lambda expression, a code pointer or an
identifier which has a function binding) and calls it with no arguments. The result
of this function is then used further on. The function is free to parse the input.

The code for the if statement, for instance, is:

```
% ***** conditional statement *****

symbolic procedure ifstat;
   begin scalar condx,condit;
    a:   condx := xread t;
         if not cursym!* eq 'then then symerr('if,t);
         condit := aconc!*(condit,list(condx,xread t));
         if not cursym!* eq 'else then nil
           else if scan() eq 'if then go to a
           else condit := aconc!*(condit,list(t,xread1 t));
         return ('cond . condit)
   end;

put('if,'stat,'ifstat);

flag ('(then else),'delim);
```

We see that ifstat is if's stat property. scan is the RLISP tokenizer (it returns identifiers, integers etc.). xread is the RLISP reader, and xread1 is as xread but does not scan a new token. aconc!* adds an element to the end of a list (using rplacd). cursym!* is the currently read token.

As an illustration, we perform our own extension of the RLISP reader, allowing to write [<expression>,...,<expression>] to enter a vector directly.

```
lisp procedure readarray;
begin
 scalar work;
a: work:=(xread t).work;
 if cursym!* = '!*comma!* then goto a
   else if cursym!* = '!] then scan()
   else symerr('![,t);
 return list('list2array ,'list.reverse work);
end;

lisp procedure list2array u;
begin
 scalar result,l;
 l:=length u;
 result:=mkvect(l-1);
 for i:=1:l do <<
  putv(result,i-1,car u);
  u:=cdr u
 >>;
```

```
 return result;
end;
```

```
put(' ! [, 'stat, 'readarray);
flag(' (!]), 'delim);
```

Yet another extension of REDUCE can be found in van Hulzen's course [9], where a case statement is described.

2.6   Standard LISP vs. Common LISP

Today, Common LISP seems to become the first real standard for LISP systems. The history of LISP is full of different dialects, each of which emphasized certain hardware possibilities and certain software properties judged to be essential by its designers. Standard LISP was introduced in 1978 as a minimal LISP standard which would allow REDUCE to run (before that, RLISP had to be implemented in each given LISP dialect). The definition of Standard LISP therefore reflects the properties which were considered necessary at that time.

In the 1980s, artificial intelligence went through a renewed wave of research activity, which caused LISP to be used more widely. The joint effect of new ideas and new people using LISP led to a demand for a new standard, which could be adhered to by many hardware and software vendors: Common LISP. This language was designed by a committee selected from academic and commercial institutions, and its specification was published in [8].

Common LISP evolved mainly from MACLISP, ZETALISP and INTERLISP, and was influenced a little by SCHEME, SPICELISP, NIL (also a LISP dialect) and Standard LISP. It is a much larger LISP system, requiring a considerable implementation effort, and Common LISP systems generally sell at quite a high price ($5000 for a workstation version is not uncommon). Luckily, there exist cheap public domain implementations, such as Kyoto Common LISP (for SUN workstations).

We survey some of the differences between Common LISP and Standard LISP:

- There are many small differences in input format and the like, which we need not explain further at this point.

- Common LISP offers, next to true integers, also a rational number data type. This is not present in Standard LISP because rationals are always handled by REDUCE itself; the internal REDUCE representation of 5/6 is (quotient 5 6) and requires no rational number data type.

- There are different types of floating-point numbers in Common LISP, which can support all IEEE floating-point formats (short, long, extended). Standard LISP only has one type of floating-point numbers.

- Common LISP has complex numbers in $a + ib$ format. These are not necessary in Standard LISP because REDUCE implements them.

- Common LISP has a function which returns the property list of an identifier. Standard LISP only allows to get an explicitly specified property, or to check whether a given flag is set (Cf. the remark under get).

- Standard LISP has vectors. Common LISP has these too, but also offers multi-dimensional arrays. In Common LISP, an array can be grown. Common LISP also lets you declare that an array will only be used to store floating-point numbers (for instance); the resulting memory usage can then be much more efficient.

- There are several data types in Common LISP which are unknown to Standard LISP: hash tables, read tables for the reader, 'random states' and packages. It is also possible to define PASCAL-like data structures. Example:
```
(defstruct lisp-course
           introduction
           lisp-basics
           lisp-overview
           extras)
```
defines a new structure type called lisp-course. Its members are general S-expressions, and may well actually be stored as its car, cadr etc. But there are new functions like lisp-course-x-introduction which extract parts of such a structure, and can also be used, with setf, to set them.

- Common LISP has a better support for function (Cf. the remark in the Standard LISP description of it).

- In Common LISP, the default binding of variables is local. So there is less difference between compiled and interpreted Common LISP. In fact, the concept of binding is split into 'scope' (where you can refer to something) and 'extent' (where it lives).

- Common LISP has no fexprs. The default action is to evaluate the arguments of a function; 'nospread' can, however, be obtained using the &rest keyword in a lambda expression. Common LISP has macros, whose argument lists can include destructuring facilities. It also has the backquote operator.

- There exist Common LISP predicates to test whether an identifier has a value (boundp) or a function value (fboundp).

- The Common LISP function setf can replace many LISP functions. Examples:

  ```
  (setf a 7)                    corresponds to  (setq a 7)
  (setf (car a) 7)              corresponds to  (rplaca a 7)
  (setf (caadr a) 7)            corresponds to  (rplaca (cadr a) 7)
  (setf (get 'a 'myprop) 15)    corresponds to  (put 'a 'myprop 15)
  ```

- cond exists in Common LISP, but there are also if, when and unless functions.

- block is a Common LISP function corresponding to progn, but which also allows return statements.

- Iteration in Common LISP can be handled using do, dolist, dotimes, which roughly correspond to the RLISP for constructs.

- The mapping functions can act on more than one list.

- Functions can return multiple values in Common LISP.

- There is a named catch / throw / unwind-protect mechanism.

- -, /, <, >, <=, >=, =, /= can act on more than two arguments.

- Common LISP has many transcendental functions (sin, cos, cosh etc.). These are not present in Standard LISP because there are REDUCE functions which compute their values.

- Common LISP has a formatted output function whose working principle is similar to the C language function printf.

Chapter 3

REDUCE ALGEBRAIC MODE

3.1 Numbers.

3.1.1 Integers

We invite you to start up a REDUCE session. How this is done depends upon the computer system but mostly it suffices to type reduce and strike the return key, perhaps after changing the current directory to the REDUCE subdirectory.

If you are successful, an acknowledgment line appears with information on the version of REDUCE you are using, followed—perhaps after a pause—by the prompt symbol 1:. You are now in the algebraic mode of REDUCE.

The instructions you want REDUCE to execute have to be entered after these numbered prompts. Each instruction must be terminated either by a semicolon or by a dollar sign. We shall explain the difference between both terminators while discussing the next examples, which are focussed on numbers.

Although the first goal of REDUCE is to manipulate symbolic expressions, it offers facilities for numeric manipulation which go far beyond these of classical calculators: the size of the numbers is only limited by the capacity of the computer memory and calculations can be carried out with exact, total precision, i.e. without any roundoff. REDUCE can handle integers, rationals, algebraic numbers, transcendental numbers, complex numbers and more—but then you must program it.

Enter, after the prompt, int1:=5**4**3; and strike the return key

*int1:=5**4**3;*

INT1 := 244140625

The typing of the carriage return started the execution of the statement: REDUCE calculated the powers involved and assigned this value to the variable int1. This assignment was caused by the symbol :=. Moreover, the result was automatically printed out and the next prompt, 2:, appeared. If, instead of the semicolon, a dollar sign had been used as terminator, the same calculation would have been executed, but the output would have been suppressed and REDUCE simply would have prompted with 2:.

From now on, and independently of what terminator was used, each time int1 appears as a variable, it will automatically be replaced by its numerical value. An example:

*int1**2;*

59604644775390625

This time, the value of the calculation has not been assigned to a variable. Should the user now wish to store this value for further use by assigning it to a variable, say square, he can give the command: saveas square; Later on, square can be used again:

sqrt(square)-int1;

0

Proceeding in the same way, we define three more variables:

*int2:=5**(4**3);*

INT2 := 542101086242752221700372640043497085571289062 5

int3:=factorial 100;

INT3 :=
93326215443944152681699238856266700490715968264381621468592963895
21759999322991560894146397615651828625369792082722375825118521091
686400000000000000000000000000

*int4:=2**67-1;*

INT4 := 147573952589676412927

Notice the brackets in the definition of int2. Arithmetical operations are carried out according to the usual precedence rules, and two consecutive alike operations, such as exponentiation, are executed from left to right.

3.1.2   Rational numbers

REDUCE offers built-in support for fractions, whose numerator and denominator can be of arbitrary size.

Examples ————————————————————————————————

rat1:=int1/int2;

$$\text{RAT1} := \frac{1}{2220446049250313080847263336181640625}$$

rat2:=6(1/2+1/3);*

RAT2 := 5

Notice that the above result for rat2 is indeed an integer and that rat1 was automatically simplified to a fraction whose numerator and denominator are relatively prime.

We check this again using the function gcd, 'greatest common divisor' (Cf. section 3.10.2):

`rat3:=int3*rat1;`

RAT3 :=

15657540749655869414778473773271793877400477942206755698088213713
50267042088016831832982472484196757096428120816469231672511916539
557785370624/37252902984619140625

`gcd(num ws, den ws);`

1

Here, we have used the built-in functions num and den which return the numerator, resp. the denominator of a rational expression (Cf. section 3.10.8).

The variable ws, 'workspace', refers to the last obtained result in the current REDUCE session. Here it represents the value of rat3. More generally, any previously obtained result can be re-used: the expression $ws(j)$ refers to the value of the instruction entered after the prompt numbered j.

There is also the function lcm (Cf. section 3.10.2), which returns the least common multiple of two or more integers.

It is possible to factorize integers into prime factors. This is done by the **factorize** function, when the switch ifactor, 'integer factorization', is on. The behavior of REDUCE is governed by a lot of (software!) switches, which toggle modes of simplification, representation etc. Generically, the switch switch is turned on by the command on switch; and turned off by the command off switch;.

If the switch ifactor is off (the default status), the function **factorize** will act usefully on algebraic expressions containing symbolic variables, but not so on numbers. But if ifactor is on, factorize will also look for the prime factors of integers.

The factorization of large integers is very time-consuming, so it can be useful to observe the time (expressed in milliseconds) taken by the calculations, by switching on time.

`on ifactor, time;`

Time: 1233 ms

The following factorization was carried out by N. F. Cole in 1903. He needed three years of Sundays.

`factorize(2^67-1);`

{193707721,761838257287}

Time: 13855 ms plus GC time: 153 ms

`off time; off ifactor;`

The last two commands switch off time and ifactor. In this way, the default status is restored.

The result of the factorization is a list, i.e. a sequence of objects enclosed by braces. The elements of a list can be extracted by specific commands (Cf. section 3.3). These commands are an algebraic mode equivalent of the corresponding Lisp commands.

3.1.3 Algebraic numbers

Using the exponentiation, we can enter fractional powers of integers:
alg1:=2**(1/2);

ALG1 := SQRT(2)

alg1**2;

2

alg2:=2**(1/3);

        1/3
ALG2 := 2

alg2**9;

8

The built-in operator sqrt, 'square root', obeys a number of simplification rules. For example, squares will always factor out:
alg3:=sqrt(18);

ALG3 := 3*SQRT(2)

But, sadly enough:
alg4:=sqrt(2 + sqrt 2) * sqrt(2 - sqrt 2);

ALG4 := SQRT(SQRT(2) + 2)*SQRT(- SQRT(2) + 2)

This can be remedied by a user-supplied simplification rule:
for all u,v let sqrt(u)*sqrt(v)=sqrt(u*v);

Such for all rules are particularly useful to express properties of mathematical functions, such as the square root, on the REDUCE operator which represents them (here, sqrt). These user-supplied rules remain active during the rest of the REDUCE session, unless they are explicitly retracted by the user.
We now indeed have
alg4;

SQRT(2)

Let us retract the rule we introduced for the square root operator:
for all u,v clear sqrt(u)*sqrt(v);

Regrettably, it is necessary that the same dummy variables u and v are used as when the rule was defined.

3.1.4 Transcendental numbers

REDUCE 'knows' about the numbers π and e (the base of the natural logarithm) and treats them symbolically. A transcendental function such as log, sin, tan generally produces a transcendental number even when applied to a rational; these values are also handled symbolically. But do not expect REDUCE to know about their relative transcendence. For instance, $\sin 2 = 2\sin 1 \cos 1$, but $\sin(2)$ will not simplify like that.

Examples ——————————————————————————————————

```
trans1:=e**7/e**2;

          5
TRANS1 := E

log(trans1);

5
```

Notice the difference:
```
trans2:=1/trans1;

          1
TRANS2 := ----
          5
          E

log(trans2);

       1
LOG(----)
       5
       E

trans3:=log 2;

TRANS3 := LOG(2)

e^trans3;

2
```

For a convex body, the minimal ratio of the squared volume to the cubed surface area is obtained for the sphere:
```
volume:=4/3*pi;

          4*PI
VOLUME := ------
            3

area:=4*pi;

AREA := 4*PI
```

```
ratio:=volume**2/area**3;
           1
RATIO := -------
          36*PI
```

3.1.5 Complex numbers

REDUCE 'knows' about the imaginary unit, which is represented by the reserved
identifier I. A built-in simplification rule replaces I**2 by (-1).

Examples ────────────────────────────────────
```
comp1:=5+sqrt(-7);
COMP1 := SQRT(7)*I + 5
comp2:=(i*sqrt(sqrt 2 + sqrt 10+2*sqrt(sqrt 5 - 5))/(2*2^(1/4)))^10;
COMP2 := 1
comp3:=1/comp1;
                1
COMP3 := ---------------
          SQRT(7)*I + 5
```

comp3 can be changed to a form where the denominator is real by switching on
rationalize. Both the numerator and denominator of comp3 are then multiplied
by the complex conjugate of the denominator.

Example ────────────────────────────────────
```
on rationalize; comp3;
  - SQRT(7)*I + 5
------------------
        32
```

The operators conj, repart and impart help manipulate complex numbers; they
respectively calculate the complex conjugate, the real part and the imaginary part
of the given complex number. Notice that when the real or imaginary part of this
number is not rational, these parts are interpreted as being complex too.

Examples ────────────────────────────────────
```
conj(3/4+7/8*i);
  - 7*I + 6
-----------
      8
```

```
repart comp1;
         1/2
 - IMPART(7   ) + 5
impart comp1;
       1/2
REPART(7   )
```

3.1.6   Floating-point numbers

Exact rational computations may consume a large amount of computer time and
memory, and sometimes a rational result, with large numbers in numerator and denominator, is difficult to interpret quantitatively.

So, if one cannot resist the temptation to see the approximated decimal value of the
exact numbers worked with, REDUCE can be set to calculate in floating point mode,
by switching on rounded.   The default precision depends on the host machine and
is shown by the command precision 0.

Examples ──

```
on rounded; precision 0;
12
rat1;
4.50359962737E-37
alg3;
 4.24264068712
ratio;
 0.00884194128288
```

If a rational number is input while rounded is on, it is automatically converted into
a floating-point number. Switching off rounded causes this floating-point number to
be reconverted to a rational one—not necessarily the original one.

Example ──

```
on rounded; number1:=2/3;
NUMBER1 := 0.666666666667
off rounded; number1;
 666666666667
 --------------
 1000000000000
```

To keep the fractional form of a rational number even when input in rounded mode, turn off the roundall switch, which is on by default.

The precision of the floating-point calculations can be freely specified by using the command precision. If the required precision is higher than the default hardware one, REDUCE changes to software-simulated floating-point arithmetic.

The printing of floating-point numbers can be influenced by the print_precision command.

Examples ———————————————————————————

precision 30;

  12

the previous precision is returned.

number1;

  0.66666666666667

it keeps its original precision.

number2:=2/3;

NUMBER2 := 0.666 66666 66666 66666 66666 66666 67

print_precision 10; number2;

0.6666666667

——

If a floating-point number is input in rounded mode, it is stored with the current precision:

precision 10; number3:=0.123456789123456789;

NUMBER3 := 0.1234567891

To keep the precision of floating-point numbers as when input, turn off the adjprec switch.

If a floating-point number is input while rounded is off, it is automatically converted into a fraction:

off rounded; number4:=123456789E-20;

```
                   123456789
NUMBER4  :=  ----------------------
             100000000000000000000
```

At any moment during an interactive REDUCE session, any previously given instruction can be called up again for re-execution. This is done by typing input(j) where the positive integer j refers to the corresponding prompt number. The instruction at the jth prompt is then re-executed, but if the working conditions have been modified (e.g. by changing switches or by introducing simplification rules), the value obtained might differ.

Clearly, you must know the number of the prompt you are interested in, before you can use ws(j) or input(j).   In this context, the commands display(n) and display($<not\ a\ number>$) can be useful. The first command results in a list of the last n instructions. The second one lists all instructions back to the beginning of the session.

3.2   Variables.

First we introduce the notion of *identifier*. An identifier has a name, which consists of an arbitrary number of alphanumeric characters. Special characters, such as *, -, = and blank, are also allowed, but each of them must be preceded by an exclamation mark '!', called an escape character.   For instance, the identifier whose name is *avant-garde* should be typed avant!-garde.

The use of such special characters reduces the chance of interference between variable names; the character * should be avoided, though, because many system identifiers contain it.

Moreover, there are quite a few reserved identifiers (such as the e, pi and i of the preceding section); they correspond to reserved variables, operators and commands. A complete list of them can be found in Appendix A of the REDUCE manual [4].

If the first character of a name is a digit, it is interpreted as a product, unless it is followed by the letter E, in which case it represents a floating-point number, if possible at all. So !23x4 is the identifier 23x4, 23x4 is the product 23*x4 and 23e4 is the number 230000.

If an arbitrary number of alphanumeric or special characters (except for double-quotes), is enclosed by double quotes, a so-called *string* is obtained.   In algebraic-mode REDUCE, strings are only used for output, in combination with the write command (Cf. 3.8), or (in some implementations) as file names (Cf. 3.13) .

Identifiers can be used to name variables, procedures, arrays, matrices and operators.

From the moment an algorithm is not completely trivial, the necessary data is not used explicitly, but is replaced by so-called variables, which are labels for data that can change.   As mentioned before, an identifier corresponds to each variable.   A mathematical expression can be assigned to it as its value, and this value itself can contain variables which either stand for themselves or represent other expressions. A variable to which a value is assigned is said to be *bound* (and the value is said to be bound to the variable), whereas a variable to which no value is assigned, is called *clear* (or *unbound*, 'still indeterminate'): it stands for itself.

In algebraic mode, a variable's value will usually be reduced to its simplest form, i.e. an expression containing only numbers and clear variables. This requires a simplification method, which proceeds in a similar way as in LISP: each variable is evaluated before its substitution in an expression etc. However, this simplification is not unique since the user disposes of many switches which he can put on or off. If no special simplification method is specified by the switches, the default method of evaluation is used. It consists of the evaluation of variables and expressions which already have a value, the expansion of the expression, the collection of like terms and their ordering. The options for controlled evaluation are treated of in section 3.11.

The command by which an expression is assigned to a variable is called an *assignment statement*. It has the form

$$<identifier> \; := \; <expression>$$

If a new expression is assigned to the same variable, the new expression supersedes the old. But what happens to another bound variable which was previously assigned to an expression containing that variable? Let us compare:

a:=c+d;

A := C + D

f:=a**2;

$$F := C^2 + 2*C*D + D^2$$

a:=u+v;

A := U + V

f;

$$C^2 + 2*C*D + D^2$$

g:=b**2;

$$G := B^2$$

b:=u+v;

B := U + V

g;

$$U^2 + 2*U*V + V^2$$

When f got its value, a was already a bound variable, so that the subsequent change of the value of a did not affect the value of f: any trace of a in f disappeared when evaluating f. On the contrary, b was unbound when assigning a value to g, so that the subsequent change of the value of b does influence the value of g.

A variable can be restored to clear status by the command

$$\texttt{clear} \; <identifier>$$

There is a second way of assigning values to a variable. It is a command of the declarative type, performed by a so-called let rule. It takes the form

$$\texttt{let} \; <identifier> \; = \; <expression>$$

Such a let rule can also be cleared using clear.

However, there are some differences with an assignment statement. Here, a bound variable appearing in < *expression* > will only be replaced by its value the first time the let rule is applied, and this evaluation process is performed every time the rule is used.

Examples ───

```
x:=a;
X := A
a:=3;
A := 3
x;
3
```

The variable x has value a, which in its turn has value 3; hence x evaluates to 3.

```
clear a; x;
A
```

x still had value a, but a is now clear, so x evaluates to a.

```
let a=5; x;
5
```

Now we changed the value of a back to 5, x (which still has value a) evaluates to 5.

──

3.3   Lists.

A list is a finite sequence of values; it can be created in two ways: the terms of the list are enumerated, separated by commas, and either enclosed by braces or given as argument (between brackets) to the function list.

Examples ───

```
l1:={a,b,y:=c,d,ee,f,g=h}
L1 := {A, B, C, D, EE, F, G=H}
l2:=list(1,1,1,{2},{3,4},{5,6,7});
L2 := {1,1,1,{2},{3,4},{5,6,7}}
l3:={};
L3 := {}
```

The value of l3 is the so-called *empty list*.

──

Two lists can be joined to form a new list by the function append. In the result, the terms of the first argument will precede those of the second argument.

Examples ──
```
append(11,12);
{A,B,C,D,EE,F,G=H,1,1,1,{2},{3,4},{5,6,7}}
append(12,13);
{1,1,1,{2},{3,4},{5,6,7}}
```

An expression can be added at the front of a list by means of the prefix function cons, or by the equivalent infix operator '.'.

Example ───
```
cons(0,12);
{0,1,1,1,{2},{3,4},{5,6,7}}
{} . 13;
{{}}
```

RLISP provides two other operations on lists, viz union and intersection, which concern lists that are sets (i.e., their elements are pairwise different). We shall 'pull up' these symbolic operators to the algebraic level, where we give them the names aunion and aintersection respectively. First, we introduce a predicate to check whether or not a given list is a set (more on boolean operators in section 3.6) and, in the same order of ideas, construct a procedure to convert a given list into a set.

```
% this is the algebraic-level declaration:
lisp operator mkset,member,setp,aunion,aintersection;

lisp procedure aunion(u,v);
if setp u and setp v then 'list.union(cdr u,cdr v)
        else rederr "at least one of the arguments is not a set";

lisp procedure aintersection(u,v);
if setp u and setp v then 'list.intersection(cdr u, cdr v)
        else rederr "at least one of the arguments is not a set";

lisp procedure setp u;
if null u then t
        else if member(car u, cdr u) then nil
                else setp cdr u;

lisp procedure mkset u;
if null u then u
        else if not member(car u,cdr u) then car u . mkset cdr u
        else mkset cdr u;
```

Examples ――――――――――――――――――――――――――――――――――――――
```
aunion(l1,l2);
```
***** at least one of the arguments is not a set
```
l4:=mkset l2;
L4 := {1,{2},{3,4},{5,6,7}}
aintersection(l1,l4);
{ }
```
――

At the symbolic level there are a number of procedures for examining and manipulating lists. Two of these are also available in algebraic mode.

- The function length returns the number of terms in the list. (More on this operator in section 3.9.)

Examples ――――――――――――――――――――――――――――――――――――――
```
    length l1;

    7

    length l3;

    0
```
――

- The reverse function returns a list which is the reverse of the argument list.

Example ―――――――――――――――――――――――――――――――――――――――
```
    reverse l1;

    {G=H,F,EE,D,C,B,A}
```
――

Extracting parts of a list is possible in various ways. If <*list*> contains m terms, the function call

$$\text{part}(<list>,n)$$

will return

- the nth term of <*list*> if $0 < n \leq m$,

- LIST if $n = 0$,

- the $(m + n + 1)^{\text{th}}$ term if $(-m) < n < 0$,

- and it will produce an error message if $n > m$ or $n < (-m)$.

For the first, second and third terms of a list one can directly use the functions first <*list*>, second <*list*> and third <*list*> respectively.

Examples ——
second 11;
B
part (11,4);
D
part (12,0);
LIST
part (11,-3);
EE
part (12,9);
***** Expression {1,{2},{3,4},{5,6,7}} does not have part 9
first 13;
***** Expression {} does not have part 1

——

The algebraic-mode equivalent of the LISP function cdr is called rest; called on a list, it returns the tail, removing the first element.

Examples ——
rest 12;
{{2},{3,4},{5,6,7}}
rest 13;
***** (list) invalid as non-empty list
rest ({} . 13);
{}

——

Another interesting procedure, but at the symbolic level, is pnth(<*list*>,n) which returns that part of <*list*> where the terms appearing in front of the nth term are deleted.
It can be lifted to the algebraic level as follows:
lisp operator apnth;
lisp procedure apnth(l,n);
if n=0 then rederr "index out of range"
        else 'list.pnth(l,n+1);

Example ———
apnth(11,3);
{C,D,EE,F,G=H}

——

If a list is given as an argument to an operator which is not one of those specified above, then a list is returned where this operator is distributed over all the elements. To avoid this, switch on listargs (for all operators globally), or declare a specific operator to be listargp.

Example ───

(More on operators and operator declarations in section 3.6.)

```
operator f,g; listargp f;
l:={1,2,3,4,5}; f l;
F({1,2,3,4,5})
g l;
{G(1),G(2),G(3),G(4),G(5)}
```

───

Finally, notice that a list value can be cleared as any expression:

```
clear l1,l2,l3;
```

3.4 Arrays

An array is a multi-dimensional table of values. The sizes in each dimension have to be specified in the array declaration, which takes the form:

$$\text{array} <identifier>(<integer\ 1>,<integer>2,\ldots)$$

where $<identifier>$ is the name of the array, and the number of arguments determines its dimension. These arguments may also be expressions that evaluate to integers.

Examples ──

```
array a(5);
array b(2,3,4);
```

—the array a is defined on the set of the 6 integers $\{0, 1, ..., 5\}$, while the array b is defined on the set of the 60 triples $\{(0,0,0), \ldots (2,3,4)\}$.

───

Once an identifier is declared to be an array it cannot be given a value anymore. However, it can be redeclared as an array with a possible change of dimension and size. Whenever an array is declared its function values are initialized at 0. Therefore, an array cannot stand for itself (as a variable could). The initial values of the array elements can be changed by an assignment statement, but not by a let rule.

Examples ──

```
a(1);
0
```

```
a(1):=pi;
A(1) := PI
let b(1,2,3)=a(1)+a(2)+a(3);
***** Substitution for 0 not allowed
```

An array can be cleared as a whole with the command

<p align="center">clear <<em>identifier</em>></p>

where <<em>identifier</em>> is the name of the array under consideration. Pay attention to the fact that clearing an individual array place results in clearing *the value* of that particular array place, if possible: the clearing of a number, for instance, leads to an error message.

Examples ————————————————————————————————————

```
clear a; b(0,0,0):=xyz;
clear b(0,0,0);
***** XYZ not found
XYZ := X*Y*Z; b(0,0,0);
X*Y*Z
clear b(0,0,0); b(0,0,0);
XYZ
```

An array can be useful when related numbers or expressions, such as the values of a mathematical function on a set of arguments, have to be stored.

Example ————————————————————————————————————

```
array c(10);
c(0):=1;
c(1):=1;
for j:=2:10 do c(j):=j*c(j-1);
```

The array c now contains the factorials from 0 to 10. Notice the for command, which will be commented on in section 3.7. This command is also required to copy arrays: an instruction like d:=c produces the error message 'Array arithmetic not defined', even if d was declared an array of the same dimension and sizes as c. The correct copying method is:

```
array d(10);
for j:=0:10 do d(j):=c(j);
```

To copy higher-dimensional arrays, multiple loops are needed.

3.5   Matrices

REDUCE's notion of a matrix coincides with the mathematical notion. The declaration statement of a matrix takes the form:

> matrix *<identifier>*(*<integer 1>*,*<integer 2>*)

or

> matrix *<identifier>*

where *<identifier>* is an identifier naming the matrix.

In the first declaration, the size of the matrix is specified (but it can still be changed by a re-declaration), *<integer 1>* indicating the number of rows and *<integer 2>* the number of columns. Both integers may be replaced by expressions that evaluate to an integer. In the second declaration, the size of the matrix has been left open and can be specified later on in the session.

This is a first difference with arrays. Another difference is that, in a matrix, rows and columns are counted from one, not from zero.

Once an identifier is declared to be a matrix, it can no longer be used as a name for a variable, list, array, etc., but it can still be re-declared as a matrix.

Just as an array, a matrix cannot stand for itself. On declaration, all entries are initialized with zero. These entries can be given another value by an assignment statement.

Examples ────────────────────────────────────

```
matrix m(3,4); m(1,1);

0

m(1,1):=1;

M(1,1)  := 1

m(2,2):=1;

M(2,2)  := 1

m(3,3):=1;

M(3,3)  := 1
```

──

The matrix m can be referred to as a whole by simply evaluating its name. It can also easily be copied to another identifier, by a simple assignment statement which provides the matrix declaration of the left-hand side.

Example ─────────────────────────────────────

```
mm:=m;
```

```
        [1   0   0   0]
        [            ]
MM  := [0   1   0   0]
        [            ]
        [0   0   1   0]
```

A matrix is cleared as a whole by the clear command, but, contrary to array clearing, an expression assigned to a matrix entry cannot be cleared (if you try this, an error message results).

Another way to declare an identifier to be a matrix, and at the same time assign values to its entries, is using the instruction

<identifier> := mat((A11, A12, A13, A14), (A21, A22, A23, A24),
                    (A31, A32, A33, A34))

where all Aij stand for expressions.

Examples

Notice the difference between the following two matrices and especially the use of the parentheses in these:

```
m1:=mat((a,b));
M1  := [A   B]
m2:=mat((a),(b));
        [A]
M2  := [ ]
        [B]
```

There are some built-in mathematical functions acting on matrices. To start with, there is matrix arithmetic, where the traditional rules with respect to the compatibility of size and the regularity of the matrices have to be respected.

Example

```
m**2;
***** Matrix mismatch
m1*m2;
[ 2     2]
[A  + B ]
m2*m1;
[ 2      ]
[A    A*B]
[       ]
[      2 ]
[A*B  B  ]
```

To compute the inverse of a matrix, just evaluate m**(-1) or 1/m. By default the Bareiss elimination method is used. The Cramer method is also implemented and is used when the switch cramer is turned on.

Example

The following code creates a Vandermonde matrix: (the mkid operator is described in section 3.9)

```
matrix vmm(3,3);
for j:=1:3 do for k:=1:3 do vmm(j,k):=mkid(a,j)**(k-1);
```

Now we invert it:

```
1/vmm;
```

```
                    A2*A3                              - A1*A3
MAT((-------------------------------- ,---------------------------------,
        2                                                    2
     A1  - A1*A2 - A1*A3 + A2*A3    A1*A2 - A1*A3 - A2  + A2*A3


             A1*A2
    ------------------------------),
                               2
     A1*A2 - A1*A3 - A2*A3 + A3


          - (A2 + A3)
    (------------------------------,
        2
     A1  - A1*A2 - A1*A3 + A2*A3


                               2    2
                             A1  - A3
    -----------------------------------------------------,
        2        2         2        2      2          2
     A1 *A2 - A1 *A3 - A1*A2  + A1*A3  + A2 *A3 - A2*A3


                               2    2
                           - A1  + A2
    -----------------------------------------------------),
        2        2         2        2      2          2
     A1 *A2 - A1 *A3 - A1*A2  + A1*A3  + A2 *A3 - A2*A3
```

$$
(\frac{1}{A1^2 - A1*A2 - A1*A3 + A2*A3}, \frac{-1}{A1*A2 - A1*A3 - A2^2 + A2*A3},
$$

$$
\frac{1}{A1*A2 - A1*A3 - A2*A3 + A3^2}))
$$

Next, the operators length, det, tp, trace and rank act on a single matrix argument and evaluate to respectively a list with the number of rows and columns, the determinant, the transpose, the trace and the rank of the considered matrix.

Examples ──

det vmm;

$$
- A1^2*A2 + A1^2*A3 + A1*A2^2 - A1*A3^2 - A2^2*A3 + A2*A3^2
$$

on factor; det vmm;

$$
- (A1 - A2)*(A1 - A3)*(A2 - A3)
$$

tp vmm;

```
[ 1    1    1 ]
[            ]
[A1   A2   A3 ]
[            ]
[ 2    2    2]
[A1   A2   A3 ]
```

trace vmm;

$$
A2 + A3^2 + 1
$$

rank vmm;

3

(It is tacitly understood that a1, a2 and a3 are pairwise different.)

──

The function nullspace constructs a basis of the kernel of the linear transformation represented by the matrix argument.

Example ──

nullspace vmm;

{}

a3:=a2; nullspace vmm;

```
{

[    1      ]
[           ]
[  - A1 - A2 ]
[------------]
[   A1*A2   ]
[           ]
[    1      ]
[  -------  ]
[   A1*A2   ]

}
```

The last function acting on matrices is mateigen, with syntax

$$\text{mateigen}(<matrix\ expression>,<identifier>)$$

It computes the eigenvalues and corresponding eigenvectors of the square matrix expression under consideration, returning a list of lists of three elements: the first element is a real square-free factor of the characteristic polynomial in the variable <identifier>, the second element is the corresponding multiplicity of this eigenvalue, and the third element is a column matrix representing the corresponding eigenvector, which may contain arbitrary constants depending upon the dimension of the eigenspace.

Example

Consider the unit matrix:
```
mu:=mat( (1, 0, 0, 0),
         (0, 1, 0, 0),
         (0, 0, 1, 0),
         (0, 0, 0, 1));
mateigen(mu,eta);
{{ETA - 1,

   4,
```

```
[ARBCOMPLEX(1)]
[             ]
[ARBCOMPLEX(2)]
[             ]
[ARBCOMPLEX(3)]
[             ]
[ARBCOMPLEX(4)]

  }}
```

A more elaborated example is the following:

Example ──

Consider a rotation matrix in three dimensional euclidean space, obtained as the composition of a rotation around the x-axis and one around the y-axis.

```
rx:=mat( (1 ,   0 ,   0 ),
         (0 ,  3/5 , 4/5),
         (0 , -4/5 , 3/5) );
ry:=mat( ( 5/13 , 0 , -12/13 ),
         ( 0    , 1 ,    0 ),
         (12/13 , 0 ,   5/13 ) );
rr:=rx*ry;
```

The product matrix rr represents the composition of the two rotations. We expect it to have the real eigenvalue $+1$ with the rotation axis as corresponding eigenvector.

```
lrr:=mateigen(rr,eta);
LRR := {{ETA - 1,


          1,


          [  - 3*ARBCOMPLEX(1) ]
          [--------------------]
          [         2          ]
          [                    ]
          [  - 2*ARBCOMPLEX(1) ]
          [                    ]
          [    ARBCOMPLEX(1)   ]

          },

                  2
          {65*ETA  - 14*ETA + 65,
```

```
1,

[   6*ARBCOMPLEX(2)*(5*ETA - 3)    ]
[   ----------------------------   ]
[           25*(ETA + 1)           ]
[                                  ]
[ 2*ARBCOMPLEX(2)*( - 5*ETA + 13)  ]
[----------------------------------]
[           25*(ETA + 1)           ]
[                                  ]
[           ARBCOMPLEX(2)          ]

}}
```

3.6 Operators

A REDUCE operator can be viewed upon as a 'function' in the mathematical language. Next to a lot of built-in operators covering the elementary functions of mathematics, the user can introduce his own mathematical functions as operators.

The built-in operators are of the prefix or of the infix type. The functional value of a prefix operator is obtained by

$$<prefix\ operator>(<arg\ 1>,<arg\ 2>,\ldots)$$

all arguments being of the required type, depending on the considered operator. For the infix operators, the functional value is obtained by the syntax

$$<unary\ infix\ operator>\ <arg>$$

or

$$<arg\ 1>\ <binary\ infix\ operator>\ <arg\ 2>$$

or still

$$<arg\ 1>\ <n\text{-}ary\ infix\ operator>\ <arg\ 2>\ <n\text{-}ary\ infix\ operator>\ <arg\ 3>\ \ldots$$
$$<arg\ n>$$

3.6.1 The built-in infix operators

For most of the infix operators, there are two identifiers, one being a special character and the other being of an alphabetic nature. They may be classified as follows:

- the assignment operator := or setq

- logical operators: or, and, not

- relational operators: = or equal, neq 'not equal', >= or geq, > or greaterp, <= or leq, < or lessp

- arithmetic operators: + or plus, - or difference (minus if unary), * or times, / or quotient (recip if unary), ** or expt (the use of the caret symbol ^ is system-dependent)

- the construction operator: . or cons.

Their alphabetic forms can also be used as prefix operators.

The assignment operator was already used in the assignment statements. The logical and relational operators will appear in the boolean expressions (Cf. section 3.7.1). The arithmetic operators need no special explanation; they were used from the beginning, in the section on numbers. The construction operator is typical for the construction of lists, which was treated of in section 3.3.

With the exception of setq and cons, all of these operators are left associative. So, for instance, a**b**c means (a**b)**c, while a.b.{c} means a.(b.{c}).

If no parentheses are used to specify the order in which a sequence of operators has to be parsed, the ordering of the above precedence list is valid: the highest precedence is attributed to cons and the lowest to setq.

3.6.2   The built-in prefix operators.

In REDUCE, there are many built-in prefix operators, some of which we already encountered, such as sqrt. Here, we shall restrict ourselves to those operators that are not studied in other sections. All of these correspond to mathematical functions.

- max and min.

  They select the greatest, resp. the smallest of a sequence of numbers. When applied to a non-numerical expression, an error message is generated.

Examples ——————————————————————————

    max(-7,0,7);

    7

    max(8);

    8

————————————————————————————————————

- abs

  It returns the absolute value of a number. When applied to an expression <*expression*>, either abs(<*expression*>) or abs(-<*expression*>) is returned.

Examples ──

```
abs(-9);

9

abs(-pi);

ABS(PI)

abs(b-a);

ABS(A - B)
```

The following mathematical functions have a simple argument, which should be a scalar expression, i.e. one not containing lists, arrays or matrices. Their values are scalar, too.

- conj, repart and impart (Cf. 3.1.5) also apply on algebraic expressions.

Example ──

```
conj(a+i*b);

 - IMPART(A)*I - IMPART(B) + REPART(A) - REPART(B)*I
```

- factorial works properly with natural numbers only

Examples ──

```
factorial 0;

1

factorial 10;

3628800
```

- nextprime returns the smallest prime number greater than its argument, which should be an integer; otherwise, an error message results.

Examples ──

```
nextprime 1111111111111111111110;

1111111111111111111111
```

```
nextprime(sqrt 2);
```

```
***** SQRT 2 invalid as integer
```

- ceiling, floor, round, fix

  If their argument is an integer, these functions return it.

  Otherwise, let r be the argument and n the integer for which $n < r < n + 1$. If r is positive, floor and fix return n, whereas ceiling returns $n + 1$. If r is negative, floor returns n, while ceiling and fix return $n + 1$.

  The function round returns the integer nearest to its argument; if the argument is of the form $n + (1/2)$, n being an integer, round returns $n + 1$ if r is positive, n if r is negative.

- exp

  The following values and properties of the exponential are known:
  ```
  exp 0;
  ```

  ```
  1
  ```

  ```
  exp(x)*exp(y)*exp(z);
  ```

  ```
   X + Y + Z
  E
  ```

  ```
  exp(i*pi/2);
  ```

  ```
  I
  ```

  ```
  exp(i*pi);
  ```

  ```
  -1
  ```

  ```
  exp(i*3*pi/2);
  ```

  ```
  -I
  ```

- log, ln, logb, log10

  The natural logarithm log has the built-in simplification rule:
  ```
  log(e**x);
  ```

  ```
  x
  ```

  The system also simplifies $\exp(\log(x))$ to x, which falsely leads to
  ```
  exp log(-1)
  ```

  ```
  -1
  ```

This can be be remedied by introducing an additional simplification rule:
for all x let exp(log(x))=expinvlog x;

The operator expinvlog is user-defined:
procedure expinvlog x;
if numberp x and x>0 then x else
*   if numberp x and x<=0 then*
*   rederr "the argument of LOG should be positive"*
*   else <<write "check that ",x,*
*   " is positive!";x>>;*

Examples

    exp(log(x+y));

    check that X + Y is positive!

    X + Y

    exp(log(2));

    2

    exp(log(-1));

    ***** the argument of LOG should be positive

- sqrt

  After simplifying its argument, the square root function (which also may be expressed by the rational power 1/2) takes the square factors out of the square root and distributes the square root function over the multiplicative non-numeric factors.

Example

    *sqrt(-60*a*b*c**2);*

    2*SQRT(B)*SQRT(A)*SQRT(15)*C*I

    where the imaginary unit I comes from the minus sign.

    Of course, this is wrong: c might be negative. It can be remedied by switching on precise:
    *on precise; sqrt(-60*a*b*c**2);*

    2*SQRT(B)*SQRT(A)*SQRT(15)*ABS(C)*I

- Other transcendental functions known to REDUCE are the circular functions
  sin, cos, tan, cot, sec and csc, the trigonometric functions sind, cosd, tand,
  cotd, secd and cscd, the hyperbolic functions sinh, cosh, tanh, coth, sech
  and csch, the inverses of these functions, atan2 (polar angle), atan2d (idem in
  degrees), erf, dilog, expint, hypot (hypothenusa function) and cbrt (cubic
  root) functions.

The system knows the exact values of sin and cos at the arguments that are
integer multiples of pi/2, pi/3, pi/4 and pi/6. Moreover, it is known that sin
is an odd function and cos an even one. Their derivatives are also known.

It is easy to introduce some useful extra functional values and properties, using
a *rule list*, to be described more fully in section 3.6.3.

```
% a predicate to recognize odd integers
procedure odp x;
if fixp x and abs(remainder(x,2))=1 then t else nil;

% a predicate to test whether an expression is
% an integer or an integer-valued arbitrary constant
procedure ffixp x;
if fixp x or arbinp x then t else nil;


procedure arbinp x;
part(x,0)=arbint;


% a strange predicate indeed
procedure strangep x;
if x=2 or x=3 or x=5 then t else
if fixp(x/2) then strangep(x/2) else nil;

circfuncrules:=
{sin(~x+pi) => -sin(x),
 cos(~x+pi) => -cos(x),
 sin(~x-pi) => -sin(x),
 cos(~x-pi) => -cos(x),

 sin(~x+pi/2) => cos(x),
 cos(~x+pi/2) => -sin(x),
 sin(pi/2-~x) => cos(x),
 cos(pi/2-~x) => sin(x),

 sin(~x+~k*pi) => (-1)**k*sin(x) when fixp k,
 cos(~x+~k*pi) => (-1)**k*cos(x) when fixp k,

 sin(~x+2*~k*pi) => sin(x) when ffixp k,
```

```
   cos(~x+2*~k*pi) => cos(x) when ffixp k,

   sin(~x+~k*pi/2) => (-1)**((k-1)/2)*cos(x) when odp k,
   sin(~k*pi/2-~x) => (-1)**((k-1)/2)*cos(x) when odp k,
   cos(~x+~k*pi/2) => (-1)**((k-1)/2)*sin(x) when odp k,
   cos(~k*pi/2-~x) => (-1)**((k-1)/2)*sin(x) when odp k,

   sin(pi/5) => sqrt(10-2*sqrt(5))/4,
   cos(pi/5) => sqrt( 6+2*sqrt(5))/4,

   sin(~x) => sqrt((1-cos(2*x))/2)
      when numberp(x/pi) and (x/pi)>0 and (x/pi)<1/6,
   cos(~x) => sqrt((1+cos(2*x))/2)
      when numberp(x/pi) and (x/pi)>0 and (x/pi)<1/6
   };
```

Examples

```
   let circfuncrules; sin(pi/24);

   SQRT( - SQRT(SQRT(3) + 2) + 2)
   --------------------------------
                 2

   cos(pi/40);

                                                         1/4
   SQRT(SQRT(SQRT(SQRT(2*SQRT(5) + 6) + 4) + 2*SQRT(2)) + 2*2   )
   -------------------------------------------------------------
                            1/8
                         2*2

   sin(7*pi-z);

   SIN(Z)
```

In many applications (such as Fourier analysis), it is necessary to express natural powers of sines and cosines as sines and cosines of multiples of the argument. These Simpson rules are now introduced by the following rule list:

```
simpson:={cos(~x)*cos(~y) => (cos(x+y)+cos(x-y))/2,
          cos(~x)*sin(~y) => (sin(x+y)-sin(x-y))/2,
          sin(~x)*sin(~y) => (cos(x-y)-cos(x+y))/2,
          cos(~x)**2      => (1+cos(2*x))/2,
          sin(~x)**2      => (1-cos(2*x))/2}$
```

Examples ———————————————————————————————

```
    let simpson;
    sin(w)**3;

        SIN(3*W) - 3*SIN(W)
    -  ---------------------
                 4


    4*cos(u)**3-3*cos(u);

    COS(3*U)

    sin(4*v)**3;

        SIN(12*V) - 3*SIN(4*V)
    -  -------------------------
                 4

    (for j:=1:4 sum mkid(a,j)*cos(j*wt))^2;

                  2
    (COS(8*WT)*A4  + 2*COS(7*WT)*A3*A4 + 2*COS(6*WT)*A2*A4


                  2
    + COS(6*WT)*A3  + 2*COS(5*WT)*A1*A4 + 2*COS(5*WT)*A2*A3


                                     2
    + 2*COS(4*WT)*A1*A3 + COS(4*WT)*A2  + 2*COS(3*WT)*A1*A2


                                     2
    + 2*COS(3*WT)*A1*A4 + COS(2*WT)*A1  + 2*COS(2*WT)*A1*A3


    + 2*COS(2*WT)*A2*A4 + 2*COS(WT)*A1*A2 + 2*COS(WT)*A2*A3


                             2    2    2    2
    + 2*COS(WT)*A3*A4 + A1  + A2  + A3  + A4 )/2
```

tan and cot are both known to be odd. Also, tan(0)=0 is known, and the derivatives.

Of sinh, cosh and tanh the parity, the value at 0 and the derivatives are known.

Of asin the parity, the value at 0 and the derivative are known. It is also
known that sin(asin x)=x, which falsely leads to:

```
sin(asin 6);
```

6

We suggest the following modifications and additions:

```
let asin 1 = pi/2, asin(-1)=-pi/2;
for all x let sin(asin x)=sininvasin x;

procedure sininvasin x;
if numberp x and abs(x)>1 then
      rederr "the argument of ASIN should belong to [-1,1]!"
   else if numberp x and abs(x)<=1 then x
   else <<write "check that ",x," belongs to [-1,1]"; x>>;
```

Examples ───

```
sin(asin 6);
```

***** the argument of ASIN should belong to [-1,1]!

```
sin(asin 1);
```

1

```
sin(asin x);
```

check that X belongs to [-1,1]

X

───

- erf.

This operator represents the error function

$$\text{erf}(x) = \frac{2}{\sqrt{\pi}} \int_0^x e^{-t^2}\, dt.$$

Its value at zero and its derivative are known.

- expint.

This operator represents the exponential integral

$$\text{Ei}(x) = \text{Pv} \int_{-\infty}^x e^t / t\, dt.$$

Its derivative is known.

- dilog.

This operator represents the dilogarithm,

$$\text{dilog } x = \int_0^x \frac{\log t}{1 - t}\, dt. \qquad (x \geq 0)$$

Its value at zero, $\pi^2/6$, and its derivative are known.

If rounded is on, the values of these mathematical functions, when given a numeric argument, will be calculated with the current floating point precision.

Examples ──────────────────────────────────────

on rounded,numval; exp 1;

2.71828182846

sin(pi/8);

0.382683432365

───

Other built-in prefix operators are discussed in section 3.9.

3.6.3 User-defined operators

Thinking of an operator as a mathematical function, it can take an arbitrary number of real or complex arguments or, in contrast with an array or a matrix, stand for itself. An identifier is declared to be an operator by the syntax

operator *<identifier>*

The number of arguments should not be specified, and no values will be initialized. Mostly, user-defined operators are used as prefix operators. However, an operator can be declared infix; its position in the precedence list of the built-in infix operators should then also be specified. The corresponding commands are infix *<identifier>* and precedence *<identifier>*,*<infix operator>*.

Examples of user-defined infix operators are the inner, outer and star operators of section 5.2.

To remove the operator character of an identifier declared operator, use a clear rule with syntax:

clear *<identifier>*

Example ——

```
operator f;
```

As long as the mathematical content of the operator f has not been explicited, f can be used with any number of arguments—even none, the expression simply remains unchanged:

```
f;
```

F

```
f();
```

F()

```
f(1);
```

F(1)

```
f(x,y,z);
```

F(X,Y,Z)

——

This is particularly useful for defining general mathematical expressions containing e.g. arbitrary constants. As an example, we introduce the general quadratic polynomial in the two variables x, y:

```
operator a;
poly2:=for j:=0:2 sum (for k:=0:(2-j) sum a(j,k)*x**j*y**k);
```

POLY2 :=

$$A(2,0)*X^2 + A(1,1)*X*Y + A(1,0)*X + A(0,2)*Y^2 + A(0,1)*Y + A(0,0)`$$

There are several possibilities to assign values to an operator. The first method is using an assignment declaration or a simple let rule to fix the values of an operator at specified arguments:

```
f(1):= u+v; let f(x,y,z)=x**y**z;
```

Any other value of the operator f remains unknown to the system:

```
f(a,b,c);
```

F(A,B,C)

If the operator f is now cleared, values such as f(1) and f(x,y,z) are apparently forgotten, but once f is re-declared as an operator, they are taken up again (which should not happen):

```
clear f; f(x,y,z);
```

Declare F operator ?   (Y or N)

Y

```
Y*Z
X
```

We see that the value $f(x,y,z)$ is indeed 'forgotten' by the system and, when we try to use f as an operator, the system offers us to redeclare it by the above question, which we answer in the affirmative way. The former value of $f(x,y,z)$ then pops up again!

A second possibility to assign values to an operator is by using a for all rule, which takes the syntax

> for all <*arg 1*>,<*arg 2*>,... let <*operator*>(...) = <*expression*>

This is particularly useful when the operator stands for a mathematical function for which a general definition is known when the arguments lie in a certain set.

Example ───────────────────────────────────────

```
for all x,y,z let f(x,y,z) = x**(y**z);
f(a,b,c);
   C
  B
A

f(2,3,4);

2417851639229258349412352
```

───────────────────────────────────────

One may want to store an expression that has just been generated, as the definition of a mathematical function. This can be achieved using the saveas command in the following way:

```
1/sqrt(x**2+y**2+z**2);

         1
--------------------
        2   2   2
  SQRT(X  + Y  + Z )

operator POT;
for all x,y,z saveas POT(x,y,z);
POT(u,v,w);

         1
--------------------
        2   2   2
  SQRT(U  + V  + W )
```

The variables appearing after 'for all' in a for all rule are dummies which do not affect the value of the ordinary variables bearing the same name. However, to clear a for all rule, *the same* dummy variables should be used in the clear command—in our last example:

```
for all x,y,z clear POT(x,y,z);
```

If several simplification rules are introduced for the same operator, by means of for all and let rules, there is no guarantee as to the order in which the rules will be applied; nevertheless a let rule is replaced by an identical rule with a different expression after the equal sign , and if two rules are in conflict, the last one prevails.

Example ——

```
operator f;
for all x,y let f(x,y)=x+y;
for all x let f(x,x)=0;
f(1,2);
```

3

```
f(1,1);
```

0

If the same rules are entered in another order, the result is different:

```
operator g;
for all x let g(x,x)=0;
for all x,y let g(x,y)=x+y;
g(1,2);
```

3

```
g(1,1);
```

2

To restrict the range of the arguments for which a simplification is valid, a for all rule with such that clause can be used. It has the syntax

> for all <*arg 1*>,<*arg 2*>,... such that <*boolean expression*> let
> <*operator*>(...) = <*expression*>

Examples ——

```
for all x,y,z such that x>0 and y>0 let f(x,y,z)=x**(y**z);
f(-1,2,3);
```

F(-1,2,3)

```
f(a,b,c);
```

***** A invalid as number

The error message is produced because the system is unable to evaluate the boolean expression, a, b and c not being numbers.

If there are several rules to be introduced, a so-called rule list can be constructed. An advantage is that the rules of this collection can be activated or deactivated simultaneously; moreover, a named rule list <*rule list name*> can be printed by the simple command <*rule list name*>;.

An example of a rule list was given in section 3.6.2, where the Simpson rules for the circular functions sin and cos were introduced:

```
let simpson; 4*cos(u)^3-3*cos(u);
```

COS(3*U)

```
clearrules simpson; 4*cos(u)^3-3*cos(u);
```

COS(U)*(4*COS(U) - 3)

If the rules in a rule list have to be constrained, a when clause should be used.

Examples ————————————————————————————————

```
operator bifac, gamma;
bifacrules:={bifac(0) =>1,
             bifac(~n)=>(for j:=1 step 2 until n product j)
                              when n>0 and remainder(n,2)=1,
             bifac(~n)=>(for j:=2 step 2 until n product j)
                              when n>0 and remainder(n,2)=0};
 gammarules:={gamma(~n) =>infinity when fixp n and n<=0,
             gamma(~n) =>factorial(n-1) when fixp n and n>0,
             gamma(1/2)=>sqrt(pi),
             gamma(~x) =>bifac(2*x-2)*sqrt(pi)/2**(x-1/2)
                         when fixp(x-1/2) and x>1/2,
             gamma(~x) =>(-2)**(1/2-x)*sqrt(pi)/bifac(-2*x)
                         when fixp(1/2-x) and x<=(-1/2) };
let bifacrules, gammarules;
for j:=-6:6 do write "gamma(",j/2,"):= ",gamma(j/2);
```

gamma(-3):= INFINITY

$$\text{gamma}\left(\frac{-5}{2}\right) := \frac{-8*\text{SQRT(PI)}}{15}$$

gamma(-2):= INFINITY

```
         - 3         4*SQRT(PI)
gamma(------):= ------------
         2               3


gamma(-1):= INFINITY


         - 1
gamma(------):=  - 2*SQRT(PI)
         2


gamma(0):= INFINITY


         1
gamma(---):= SQRT(PI)
         2


gamma(1):= 1


         3         SQRT(PI)
gamma(---):= ----------
         2              2


gamma(2):= 1


         5         3*SQRT(PI)
gamma(---):= ------------
         2              4


gamma(3):= 2
```

If the mathematical expression to be assigned to an operator is more complicated than a traditional closed-form definition (it might include branch selection or iteration), one can use a fourth method of assigning values to operators, the so-called *procedure declaration*. In a procedure declaration, the arguments have to be specified from the beginning. It has the syntax

[algebraic] procedure *<identifier>* (*<arg 1>*,*<arg 2>*,...);
begin
<procedure body>
end;

The word **algebraic** may be omitted. The procedure body contains the commands to be executed when calculating the value for the specified arguments. It also contains the specific command for returning such a value, which has the syntax:

$$\texttt{return} \ <expression>$$

This procedure body can be a group statement, but usually it is a compound statement or **begin** ... **end** block (Cf. section 3.7.2 for information on group and compound statements).

In this compound statement, parameters may appear to which only temporary values are assigned. They are called local variables, and have to be declared as such right after the beginning of the procedure body. This declaration has the syntax

$$\texttt{scalar} \ <identifier\ 1>,<identifier\ 2>,\ldots$$

These scalar local variables cannot be redeclared as arrays or matrices, but their values can be lists. As the name suggests, the local variables do not exist outside the procedure body. Any array, matrix or operator declared in the compound statement is automatically global in scope and carries its values, if any, outside the procedure body. Similarly, if an array, matrix or operator is cleared within a procedure body, it is cleared globally.

3.7   Expressions and statements

3.7.1   Expressions.

A REDUCE expression is a sequence of alphanumeric characters restricted to a certain syntax. Before a candidate expression is evaluated, it is syntactically checked. A syntactical error results in an appropriate error message. There are several types of REDUCE expressions, some of which we have already encountered in the preceding sections.

- Integer expressions.

  An integer expression is an expression which will evaluate to an integer. They are of interest in boolean expressions.

- Scalar expressions.

  A scalar expression is an expression which will evaluate to a real or complex number when replacing all variables by their actual or future value. Also elements of an array, a list or a matrix and values of an operator may occur in it.

  As was already mentioned in section 3.7.1, the evaluation process of a scalar expression is subject to the default or user-determined switch settings.

- Boolean expressions.

  A boolean expression evaluates to a truth value, i.e. 'yes' or 'no', or in LISP: t, 'true' or nil, 'false' (both identifiers are also reserved in algebraic-mode REDUCE).

Such a boolean value cannot be assigned to an algebraic-mode variable.

It has to appear directly within an if, for, while or repeat statement, or in the such that clause of a for all rule. Boolean expressions can have one of the following two syntaxes:

<expression> <relational operator> <expression>

or

<boolean function> (...)

or can consist of boolean subexpressions connected by logical operators:

<boolean expression 1> <logical operator> <boolean expression 2>

The relational and logical operators were already enumerated in section 3.6.

The infix operators < and > only accept integers, rationals and floating-point numbers as arguments. The infix operators = and neq check whether or not the difference of both arguments is zero. This might be influenced by the current switch settings.

With the logical operator and the boolean expressions are evaluated from left to right until a false expression is found, otherwise the result is 'true'. With the logical operator or the evaluation of the boolean expressions from left to right stops as soon as a true expression is found. Otherwise, the result is 'false'.

The logical operator not changes 'true' into 'false' and vice-versa.

The built-in boolean functions are the following:

- numberp(x)
  It is 'true' iff x is an integer, a rational number or a floating-point number.
- fixp(x)
  It is 'true' iff x is an integer.
- evenp(x)
  It is 'true' iff x is an even integer
- freeof(u,v)
  It is 'true' if the expression u does not contain the kernel v. (For the notion of 'kernel', see section 3.7.1.)
- ordp(u,v)
  It is t (i.e., true) if, in the present ordering of identifiers, u is ordered ahead of v (more on the ordering and the way the user can influence it in section 3.11).
- primep(u)
  It is t if u is prime.

Examples ──

```
operator f;
for all x,y,z such that numberp x and numberp y and x>0 and y>0
              let f(x,y,z)=x**(y**z);
f(2,3,4);

2417851639229258349412352

f(-1,2,3);

F(-1,2,3)

f(a,b,c);

F(A,B,C)
```

──

Boolean functions can also be user-made, even at the algebraic level. If an algebraic expression occurs where a boolean one is to be expected, then it is interpreted as 'false' if it evaluates to 0 and as 'true' otherwise.

Example ──

```
procedure naturalp x;
if fixp x and x>=0 then 1 else 0;
```

This new boolean function naturalp can now be used in this (rather inefficient!) recursive definition of the Fibonacci sequence.

```
procedure fibo(n);
if not naturalp n then
  rederr "the argument must be a natural number"
else if n=0 then 0
else if n=1 then 1 else fibo(n-2)+fibo(n-1);
```

The following behavior is then obtained:

```
fibo(10);

55

fibo(-5);

***** the argument must be a natural number
```

──

- Equations.

  An equation is an expression with the syntax

$$< expression > \; = \; < expression >$$

This is a boolean expression, but can also serve as an argument to operators such as solve (Cf. section 3.9.3), or appear as a value, e.g. as an element of a list.

In an equation, the left-hand side remains unchanged, while the expression at the right-hand side is simplified. If the left-hand side too has to be simplified, then evallhseqp should be switched on.

Example ───

```
equ:=(a+b)^3+(a-b)^3=(u+v)^2+(u-v)^2;

         3         3   2     2
EQU := (A + B)  + (A - B) =2*U  + 2*V

on evallhseqp; equ;

   3        2 2     2
2*A  + 6*A*B =2*U  + 2*V
```

──

The functions lhs and rhs act on an equation and return the left resp. right-hand side of their argument.

Examples ──

```
a:=3*x-4*y=z/2;

                Z
A :=  - 3*X + 4*Y=---
                2

lhs a - rhs a;

  6*X - 8*Y + Z
  ---------------
          2
```

──

- Conditional expressions.

  A conditional expression has the syntax

    if *<boolean expression>* then *<expression 1>* [else *<expression 2>*]

If the boolean expression evaluates to 'true', the conditional expression evaluates to the value of $<expression\ 1>$. If the boolean expression evaluates to 'false', it evaluates to $<expression\ 2>$ if present, otherwise the value of the conditional expression is interpreted as zero.

An example of a conditional expression can be found in the above definition of the function fibo.

3.7.2 Statements.

In the previous sections we already met several statements such as the assignment in section 3.2 and the conditional in section 3.7.1. We now give an overview of the different types of REDUCE statements.

A REDUCE statement combines identifiers, reserved words, special characters and expressions. A statement that is followed by a terminator—either a semicolon or a dollar sign—is a command for REDUCE. These commands are the building blocks of a REDUCE session.

- Assignment statements.

  A general assignment statement has the syntax

  $$<expression\ 1> := <expression\ 2>$$

  It may appear as a command, but also in a group statement or in a procedure.

  At the left-hand side, $<expression\ 1>$ can be an identifier, a specified element of an array or a matrix, or the value of an operator at specified arguments. Even more complicated expressions may appear at the left-hand side (Cf. section 3.2).

  The assignment statement will evaluate $<expression\ 2>$ and assign the resulting value to $<expression\ 1>$, if possible. Indeed, if $<expression\ 2>$ evaluates to a matrix, it can only be assigned to an identifier which then is declared as a matrix.

  If the assignment statement is turned into a REDUCE command by terminating it by a semicolon, it is printed as

  $$<expression\ 1> := <value\ of\ expression\ 2>$$

  If a dollar sign is used as terminator, the assignment action will take place, but nothing will be printed.

  A set statement can be used instead of an assignment statement if the expression at the left-hand side also has to be evaluated. The syntax then is

  $$set(<expression\ 1>,<expression\ 2>);$$

- Group statements.

  A group statement is a finite sequence of statements, separated by semicolons
  or dollar signs (here it is irrelevant which of both is used) and enclosed between
  the symbols << and >>. The statements in a group statement are executed
  one after another from left to right. If the last statement in the group has a
  value then the group statement evaluates to that value. However, if the group
  statement has a terminator at the end (in front of >>), it evaluates to zero. A
  group statement can be used wherever an expression can occur.

Examples ————————————————————————————————————

```
a:=<<b:=c;d:=ee>>;
```

A := EE

```
a;
```

EE

```
b;
```

C

```
c;
```

C

```
d;
```

EE

```
a:=<<b:=c;d:=ee;>>;
```

(no output)
```
a;
```

(no output)
```
b;
```

C

```
c;
```

C

```
d;
```

EE

- Conditional statements.

  A conditional statement has the syntax

  > if *<boolean expression>* then *<statement 1>* [else *<statement 2>*]

  If the boolean expression evaluates to 'true', *<statement 1>* is executed, otherwise *<statement 2>* is executed (if present). Both *<statement 1>* and *<statement 2>* may themselves be conditional statements.

Example ————————————————————————————

```
matrix m(4,4);
for j:=1:4 do
  for k:=1:4 do
    if j=k then
      if j<3 then
        m(j,k):=1
      else
        m(j,k):=-1
    else
      if j>k then
        m(j,k):=2;
m;
```

```
[1   0   0    0 ]
[                ]
[2   1   0    0 ]
[                ]
[2   2  -1    0 ]
[                ]
[2   2   2   -1]
```

- for statements.

  The REDUCE for loop expresses an iteration and is particularly suited for a program loop where the number of repetitions is known in advance. It has the following syntax:

  > for *<control variable>* := *<integer 1>* step *<integer 3>*
  > until *<integer 2>*

  or

for each <*control variable*> in <*list*>

followed by the action: either

do <*statement*>

or

[sum | product | collect | join] <*expression*>

In the first case, the action takes place as long as the value of the control variable, which increases by the amount <*integer 3*>, stays between <*integer 1*> and <*integer 2*>. For convenience, it is allowed to substitute a colon for the construction step 1 until.

In the second case, the action is performed for <*control variable*> equal to each of the elements of <*list*> consecutively.

From the moment the value of the control variable exceeds <*integer 2*>, or <*list*> is exhausted, the loop is stopped.

If the action is a do action, the statement is executed repeatedly and no value is assigned to the for statement. Otherwise, <*expression*> is first evaluated (if <*control variable*> appears explicitly in this expression, its temporary value is used in the evaluation of <*expression*>). The sum action causes these values to be added up; the product action will multiply them; the collect action will build a list storing the values in order; the join action only makes sense if <*expression*> always evaluates to a list: it will then join these list values to make the result.

In these four cases, the result of the action is the value assigned to the for statement. Notice that in a for statement the control variable is automatically declared as a local variable, so the loop does not affect its value outside the loop. Therefore, i can be used as a counter, without fear of interference with the system rule I**2=-1.

Examples ──

```
for j:=1:10 sum j**2;
```
385

```
l:=for i:=1:10 collect i;
```
L := {1,2,3,4,5,6,7,8,9,10}

```
for each x in l collect x**2;
```
{1,4,9,16,25,36,49,64,81,100}

- while statements.

  A while statement has the syntax:

  $$\texttt{while } <boolean\ expression> \texttt{ do } <statement>$$

  and is particularly suited for a program loop where the number of iterations is not exactly known in advance.

  This is e.g. the case if a series is summed for as long as the terms are bigger than a fixed quantity.

Example ————————————————————————————————————

The following code computes an approximation to $\sum_{n=1}^{\infty} \frac{1}{n^2}$:

```
series1:=0;
term:=1;
while (term - 1/1000) > 0 do <<
        series1:=series1 + term;
        mret:=1/sqrt(term);
        term:=(mret+1)**(-2) >>;
series1;
```

```
 84097188293211117760031704609
-------------------------------
 52130965220736832332302400000
```

In a while statement, the boolean expression is tested before the statement following do is executed. As soon as it evaluates to 'false', the while loop is terminated.

The repeat statement shows resemblance with the while statement, in the sense that it induces an iteration to take place depending on a certain condition. It syntax is

$$\texttt{repeat } <statement> \texttt{ until } <boolean\ expression>$$

In contrast with the while statement, the boolean expression is tested after the execution of the statement, and if it evaluates to 'true', the action is stopped.

Example ——

The same series can be approximately summed as follows:
```
series2:=0;
term:=1;
repeat <<
        series2:=series2+term;
        mret:=1/sqrt(term);
        term:=(mret+1)**(-2) >>
        until (term - 1/1000) <= 0;
series2;
```

$$\frac{8409718829321111776031704609}{5213096522073683233230240000}$$

——

- Compound statements.

The two constructions for obtaining the sum of the inverses of the squares of the natural numbers up to a certain order, cited in the preceding subsection, are examples of bad programming. Indeed, an auxiliary variable mret was introduced in both of them as a global variable carrying its last value outside the calculation of the series.

It may also happen (e.g. in the definition of a mathematical function by means of a procedure) that a branch selection or iteration is needed which cannot be expressed conveniently using while or repeat statements.

A finite sequence of statements that have to be executed to obtain a final result can be enclosed between the reserved words begin and end, which serve as delimiters rather than as commands (so a semicolon is never needed after the word begin, nor just before the word end). The statements themselves are separated by terminators.

The result is called a compound statement or begin ... end block. It is more general than a group statement. We already encountered an example of a compound statement when discussing the definition of user-made functions by means of procedures.

The values of intermediate results can be assigned to local variables. Those values cease to exist outside the compound statement, and there is no conflict between a local variable and a global variable outside the compound statement bearing the same name.

This is of much importance when working with large programs. Moreover the storage space occupied by the values of the local variables is freed on leaving the compound statement.

The declaration of the local variables has to be done immediately after the word begin, and has the syntax:

$$\texttt{scalar} \; <identifier> \; [,\ldots]$$

or

$$\texttt{integer} \; <identifier> \; [,\ldots]$$

or

$$\texttt{real} \; <identifier> \; [,\ldots]$$

where scalar, real and integer refer to the types of the local variables.

All local variables are initialized to zero, so they never are clear.

A compound statement only has a value if it contains a return statement, which consists of the command return followed by an expression.

If several return commands appear in a compound statement (as a part of conditional statements), the first return command reached is executed and the rest of the compound statement is neglected.

Example ───────────────────────────────────────

Using a compound statement, the summation previously described becomes:

```
begin
      scalar series,term,mret;
      term:=1;
      while (term-1/1000) > 0 do
              << series:=series+term;
                 mret:=1/sqrt(term);
                 term:=(mret+1)**(-2) >>;
      return series
end;

   840971882932111177603170460 9
   --------------------------------
   521309652207368323323024000 0
```

3.8 Commands.

In section 3.7.2, we already defined a REDUCE program as a finite sequence of commands, a command being a statement followed by a terminator. The aim of this section is to give an overview of the most important built-in commands. We also refer to the REDUCE User's Manual [4]

3.8.1 bye and quit.

The bye command terminates REDUCE, clearing all objects and rules, closing the open files and returning to the computer system's shell program: the whole REDUCE session is destroyed.

The result of the quit command depends upon the implementation. It may be synonymous to bye, but it may also have the meaning of *interrupting* the execution of REDUCE to return to the computer's shell program, while saving the full contents of the present REDUCE session for re-use later on.

3.8.2 define

It sometimes happens that the user wants to modify an element of the REDUCE syntax (e.g. an electrical engineer preferring to use J instead of I for the imaginary unit). The define command is used to supply a new name for a number, an expression, an identifier representing a variable, list, array, matrix, operator, or even a command. Its syntax is

$$\text{define } <identifier> = <expression>$$

where $<expression>$ stands for any of the above-mentioned REDUCE objects. This replacement is executed before any simplification or evaluation and remains in effect during the whole REDUCE session.

Examples ———————————————————————
```
define & = sqrt;
define xx = & y + & z;
xx**2;
2*SQRT(Z)*SQRT(Y) + Y + Z
```
———————————————————————————————

3.8.3 write

When an expression or an assignment statement is terminated by a semicolon, its value is output, normally to the screen. It may, however, be useful to print intermediate results from within for statements, group and compound statements, procedures and the like. This is made possible by the write command. Its syntax is

$$\text{write } [\ <expression> \ | \ <assignment> \ | \ <string> \ ]$$

In the first case, the expression is evaluated and its value is printed:

Example ————————————————————————
```
for j:=1:3 do write j**3;
1
```

8

27

In the second case, the expression at the right-hand side of the assignment is evaluated and is printed out after the symbol at the left-hand side, followed by the := sign. This is particularly useful when the left-hand side is an element of an array or a matrix, or an operator value with a control variable as argument, e.g. in a for statement:

Example —————————————————————————————

```
array cube(3); for j:=1:3 do write cube(j):=j**3;
CUBE(1)  := 1

CUBE(2)  := 8

CUBE(3)  := 27
```

In the third case, the string, enclosed by double-quotes, is simply printed out. These three possibilities for a write command may be mixed freely, the subsequent objects to be printed out being separated by commas. Whenever a write command occurs, a new line is taken in the output, the subsequent expressions being printed adjacent to each other with splitting over line boundaries if necessary. An empty string will correspond to an empty line.

3.8.4   Substitution commands

Substitution of variables in previously obtained expressions is an important tool in mathematics. In REDUCE, substitutions can be performed in three ways: by means of the prefix operator sub, the infix operator where, and by let rules.

sub

The syntax of the sub operator is:

$$\text{sub}(<identifier\ 1>=<expression\ 1>,\ldots,<scalar\ expression>)$$

It is executed in several steps:

- first, all expressions are evaluated, using all assignments and rules valid before the sub command;

- next, each $<identifier\ i>$ occurring in $<scalar\ expression>$ is replaced by the corresponding $<expression\ i>$; this replacement is done in parallel, which means that a $<identifier\ i>$ occurring in an $<expression\ j>$ is not affected by the substitution;

- finally, the substituted scalar expression is re-evaluated.

It is important to notice that the sub command only allows substitution in scalar expressions, not in lists, arrays or matrices. However, <*scalar expression*> need not to be explicit, it may be replaced by a variable or an expression having that value. During the substitution, no assignments take place: the value of each <*identifier i*> is unchanged when the substitution is carried out.

Examples ——————————————————————————————————

```
u:=x**2+y**2;

      2   2
U := X  + Y
v:=sub(x=a+b,y=a-b,u);

          2   2
V := 2*(A  + B )
x;
X
y;
Y
u;
 2   2
X  + Y
```

If it is desirable to make the effect of a substitution in a variable permanent, assign the value of the substitution to that variable:

```
u:=sub(x=a+b,y=a-b,u);

          2   2
U := 2*(A  + B )
```

A final remark about this built-in sub command: it also allows the substitution for operators. We illustrate this:

Example ——————————————————————————————————

```
operator a,b;
poly2:=for j:=0:2 sum (for k:=0:(2-j) sum a(j,k)*x**j*y**k);
                2                                    2
POLY2 := A(2,0)*X  + A(1,1)*X*Y + A(1,0)*X + A(0,2)*Y  + A(0,1)*Y

         + A(0,0)
```

```
poly3:=sub(a=b,poly2);
                  2                                    2
POLY3 := B(2,0)*X  + B(1,1)*X*Y + B(1,0)*X + B(0,2)*Y  + B(0,1)*Y

        + B(0,0)
```

---

where

The infix operator where has the syntax

$$<expression> \text{ where } <rule\ list>$$

where *<rule list>* may consist of only one rule, in which case the braces are not necessary. The rules contained in *<rule list>* are only applied on the expression at the left-hand side, and neither affect any other expression nor other let rules.

Examples ———————————————————————

```
operator f,g;
ex1:=f(x)**2+g(y)**2; ex2:=f(u)**2+g(v)**2;
rl1:={f(x)=>a+b,g(y)=>a-b}; rl2:={f(~x)=>a+b,g(~x)=>a-b};
ex1 where rl1;
     2    2
2*(A  + B )
ex2 where rl1;
     2      2
F(U)  + G(V)
ex2 where rl2;
     2    2
2*(A  + B )
```

This where operator can also be used for substitution in lists.

Example ———————————————————————

```
l:={a,b,c,d}$ l where a=>s+t;
{S + T,B,C,D}
```

The where operator can also be used in symbolic mode expressions, where it generates a lambda construct.

Notice that the infix operator where has a lower precedence than all other infix operators.

let rules

We already encountered let rules as assignment statements, and also when discussing the definition of operators as mathematical functions. In the latter case, they had the form

$$\text{for all } \ldots \text{ let } \ldots$$

or

$$\text{for all } \ldots \text{ such that } \ldots \text{ let } \ldots$$

A simple let rule has the syntax:

$$\text{let } <expression\ 1> = <expression\ 2>$$

For appropriate choices of both expressions, this let rule can act as an assignment to various objects such as variables, lists, arrays, matrices and functional values of an operator.

We now discuss the action of this simple let rule as a substitution.

- The $<expression\ 2>$ must be a scalar expression. It is evaluated before the substitution action takes place.

- The $<expression\ 1>$ may be a variable, an array element or a matrix element, or the functional value of an prefix or infix operator at specified arguments.

- Once the let rule is entered, it is valid throughout the rest of the current REDUCE session.

- let x=a+b,y=a-b will cause any x and any y to be replaced by (a+b) and (a-b) respectively. In fact these are two assignments.

- operator f; let f(x,y,z) = x**y**z;

  will cause f(x,y,z) to be replaced by x**y**z, but it will leave any other functional value of the operator f unchanged.

- array a(9); let a(0) = 1;

  will lead to an error message since, by the 'instantaneous evaluation property' of an array (and also of a matrix), a(0) stands for 0 and no substitutions for numbers are allowed.

  To define the elements of an array or a matrix, a proper assignment statement must be used.

- let x+y=z;

will cause any x to be replaced by z-y, and so indirectly x+y to be replaced by z. The let rule

let y+x=z;

will have the same effect—it will not cause y to be replaced by z-x. This is due to the internal ordering of the identifiers (Cf. also 3.11.5).

• let a*b**2=c;

will cause the replacement of a*b**2 by c in any expression which is a multiple of a*b**2. So a**2*b**3*c**4 will be changed by this rule to a*b*c**5. If the substitution is to be executed only when the product appears in its literal form, a match rule should be used instead of a let rule. So

match a*b**2=c

will not affect the expression a**2*b**4*c**5, but it will simplify the expression x*a*b**2*c*y to c**2*x*y.

• let x**9=0;

will cause all powers of x, greater than or equal to 9, to be replaced by zero.

What is special to this last substitution is that it is executed during the manipulation of the polynomials in the variable x, rather than only being applied to their final, simplified, values. This is particularly useful when handling polynomials as truncated power series approximations of mathematical functions: not only unnecessary computations are avoided by neglecting superfluous powers of x, but essentially incorrect terms of higher order are excluded.

Example ———————————————————————————————————————

```
operator a;
poly1:=for j:=0:3 sum a(j)*x**j;
let x**7=0;
poly1**3;
```

$$3*A(3)^2*A(0)*X^6 + 6*A(3)*A(2)*A(1)*X^6 + 6*A(3)*A(2)*A(0)*X^5$$

$$+ 3*A(3)*A(1)^2*X^5 + 6*A(3)*A(1)*A(0)*X^4 + 3*A(3)*A(0)^2*X^3 + A(2)^3*X^6$$

$$+ 3*A(2)^2*A(1)*X^5 + 3*A(2)^2*A(0)*X^4 + 3*A(2)*A(1)^2*X^4$$

$$+ 6*A(2)*A(1)*A(0)*X^3 + 3*A(2)*A(0)^2*X^2 + A(1)^3*X^3 + 3*A(1)^2*A(0)*X^2$$

$$+ 3*A(1)*A(0)^2*X + A(0)^3$$

To handle approximating polynomials in several variables—say, two, x and y—, a different approach is necessary, because not only specific powers of x and y are to be neglected but also weighted product combinations of powers of x with powers of y up to a certain total weight. The commands:

weight x=<*integer 1*>
weight y=<*integer 2*>
wtlevel <*integer 3*>

will cause a term to disappear whenever the sum of <*integer 1*> times the power of x and <*integer 2*> times the power of y exceeds <*integer 3*>. The default value of the weight level wtlevel is set to one, and variables not mentioned in a weight declaration are of weight zero.

Examples ───

poly:=x**2+y**2+x+y+1;

$$POLY := X^2 + X + Y^2 + Y + 1$$

popoly:=poly2**2;

$$POPOLY := X^6 + 2*X^4 + 2*X^3*Y^3 + 2*X^3*Y^3 + 2*X^3 + X^3 + 2*X^3*Y + 2*X*Y^3$$

$$+ 2*X + Y^6 + 2*Y^4 + 2*Y^3 + Y^2 + 2*Y + 1$$

weight x=2; weight y=1; wtlevel 4; popoly;

$$X^2 + 2*X*Y + 2*X + 2*Y^4 + 2*Y^3 + Y^2 + 2*Y + 1$$

wtlevel 1; popoly;

$$2*Y + 1$$

wtlevel 0; popoly;

$$2*Y + 1$$

──

The let rule allows even more general substitutions than those mentioned above. The substituting expression <*expression 1*> is first simplified by collecting like terms and putting the identifiers in the default order. No substitutions on <*expression 1*> are performed, except for the 'instantaneous evaluation' of array and matrix elements. Two special cases may occur:

- if the left-hand side expression shows the multiplication operator at top-level, then the constants are moved to the right-hand side expression. So,
  `let 5*x*y*z = u+v`

  will cause x*y*z to be replaced by (u+v)/5 in any expression which is a multiple of x*y*z. The same holds for a more complicated product such as
  `operator f;`
  `let f(x)*f(y)*f(z)=f(x+y+z);`
  `f(x);`

  F(X)

  `f(x)*f(y);`

  F(X)*F(Y)

  `f(x)*f(y)*f(z);`

  F(X+Y+Z)

  `f(x)**3*f(y)**4*f(z)**6;`

$$F(X + Y + Z)^3 *F(Y)*F(Z)^3$$

- if the left-hand side expression shows the operators +, - or / at top level, then only the first term (after ordering !) is kept at the left-hand side, while the rest is moved to the right-hand side. So
  `let b-c+a=x;`

  will result into the replacement of a by x-b+c, and, in the same order of ideas,
  `let f(x)+f(y)+f(z) = f(x*y*z)`

  will cause f(x) to be replaced by f(x*y*z)-f(y)-f(z).

A final word about how to clear a let rule. We already discussed the clearing of assignments and of (the elements of) an array or a matrix. To undo the rule

$$\text{let } <expression\ 1> = <expression\ 2>$$

use the command

$$\text{clear } <expression\ 1>$$

If the system has itself modified the let rule, a simpler clear command may be sufficient. For example, the rule
`let f(x)+f(y)+f(z)=f(x*y*z)`
is cleared by the command
`clear f(x)`

3.9   Special prefix operators.

3.9.1   Differentiation

Assume that the value of f is an algebraic expression in the variables x1, x2 and x3, which are themselves unbound variables. This expression can be viewed upon as a mathematical function of the three real variables x1, x2 and x3. The partial derivative of order n1+n2+n3 of the function f with respect to x1 (n1 times), to x2 (n2 times) and x3 (n3 times) is then given by

df(f,x1,n1,x2,n2,x3,n3)

For a derivative of the first order, the number of derivations (i.e. 1) may be omitted.

Example ──

Consider the Newtonian potential V in euclidean three-space: we shall prove that it is harmonic in the complement of the origin.

```
operator x;
V:= 1/ sqrt(x(1)**2+x(2)**2+x(3)**2); df(V,x(1),2);
```

```
         2      2        2         2       2       2      4
 - (X(3)  + X(2)  - 2*X(1) )/(SQRT(X(3)  + X(2)  + X(1) )*(X(3)  + 2*


         2     2        2     2      4        2     2      4
      X(3) *X(2)  + 2*X(3) *X(1)  + X(2)  + 2*X(2) *X(1)  + X(1) ))
```

That was one of the derivatives. We must add them to obtain the laplacian:

```
procedure laplacian(f,n);
for j:=1:n sum df(f,x(j),2);
```

Now we can apply it:

```
laplacian(V,3);
```

```
0
```

──

In the evaluation of df(f,x), first f and x are evaluated. Assuming that x is a clear variable, the expression assigned to f is differentiated with respect to x, according to the known differentiation rules.

These include the differentiation of the built-in standard elementary functions and the chain rule. If the value of f is independent of x, this derivative equals zero.

However, f might be a clear variable, and stand for itself. Then df(f,x) would automatically evaluate to zero, unless f had been declared to be an unspecified function of x, in other words, to depend upon x. This declaration has the syntax:

depend f,x,y,z;

—meaning that f is to be considered as a function of the variables x, y and z.

A dependence introduced in such a way can be cleared by the command nodepend, which has the syntax

nodepend f,y,z;

—meaning that f should no longer be seen as dependent on y and z (but still on x).

Example

Let us show that REDUCE applies the chain rule for differentiation:
`f:=sin g; depend g,x; df(f,x,3);`

$$DF(G,X,3)*COS(G) - 3*DF(G,X,2)*DF(G,X)*SIN(G) - DF(G,X)^3*COS(G)$$

However, the chain rule involving only unspecified functions is not applied by RE-DUCE:
`clear f; depend f,g; depend g,x; df(f,x);`
`DF(F,X)`
instead of `df(f,g)*df(g,x)`.

No dependence declaration is needed for the differentiation of an operator f when, after simplification, the independent variable occurs in some arguments to f.

Example

`operator f; df(f,x);`
`0`
`df(f(x,y,z),x);`
`DF(F(X,Y,Z),X)`

Finally, it is also possible to introduce rules for the differentiation of user-defined operators, by means of a particular let-rule of the form:

$$\texttt{for all} \ <identifier\ 1>,\dots\ \texttt{let}$$
$$\texttt{df}(<operator>(<identifier\ 1>,\dots)=<expression>$$

Examples

`operator f,h;`
`for all x,y,z let df(f(x,y,z),x,2)=h(x,y,z);`
`df(f(x,y,z),x);`
`DF(F(X,Y,Z),X)`
`df(f(x,y,z),x,2);`
`H(X,Y,Z);`
`df(f(x,y,z),x,3,y,2,z);`
`DF(H(X,Y,Z),X,Y,2,Z)`

3.9.2   Finding primitives

If the value of f is an expression containing the unbound variable x, f can be interpreted as a mathematical function of the real variable x. If this function possesses a suitable closed-form primitive (an indefinite integral) in a certain open subset of the real axis, it can be computed by evaluating

`int(f,x)`

Using a combination of the Risch-Norman algorithm and pattern matching, this `int` operator searches for a primitive in closed form, in terms of the built-in elementary functions. If it is not successful, it will return the input, possibly in an alternative form containing other (and perhaps more complicated) primitives. It can also be useful to introduce closed-form primitives that are not found or expressed by the system in a too complicated form.

Examples ──

`df(asin(x),x);`

```
            2
   SQRT( - X  + 1)
 - -----------------
          2
         X  - 1
```

`int(ws,x);`

```
                 2
         SQRT( - X  + 1)
 - INT(-----------------,X)
              2
             X  - 1
```

We add a `for all` rule to extend REDUCE's knowledge a little...

`for all x,a let int(1/sqrt(a-x**2),x)=asin(x/sqrt(a));`

... and verify:

`on precise; int(1/sqrt(alfa**2-x**2),x);`

```
          X
ASIN(-----------)
        ABS(ALFA)
```

Another example:

`df(dilog(x),x);`

```
   - LOG(X)
 -----------
    X - 1
```

```
int(ws,x);
          LOG(X)                    2
  - 2*INT(--------,X) - LOG(X)
             2
          X  - X
------------------------------
                 2
```

Again, we extend the integration operator:

```
for all x let int(log(x)/(x^2-x),x)=-dilog(x)-log(x)^2/2;
w:=int(log(x)/(1-x),x);
W := DILOG(X)
```

Even when the primitive cannot be expressed in terms of elementary functions, RE-DUCE knows that df and int are inverse operators.

Example

```
int(sin x/x,x);
      SIN(X)
INT(--------,X)
        X
df(ws,x);
  SIN(X)
--------
    X
```

Even if f is declared as operator, int(f,x) evaluates to f*x unless f has a value. But if the integration variable appears among the arguments to f, the input is returned: int(f(x),x), for instance, evaluates to int(f(x),x). For the df operator a similar problem could be solved using a depend declaration; this does not apply here, for if we enter depend f,x; the evaluation of int(f,x) leads to an error message.

More on the subject of symbolic integration can be found in e.g. [3].

As a final example, we construct a procedure to compute the Fourier series expansion of a piecewise continuous function in the interval $[-\pi, \pi]$, up to a certain rank.

First, we define a procedure to compute the integral of a continuous function over a given interval.

```
procedure integral(function,interval,x); begin
  scalar u,a,b;
  u:=int(function,x);
  a:=first interval;
  b:=second interval;
  return sub(x=b,u)-sub(x=a,u) end;
```

Example ───

```
integral(sin(x),{0,pi},x);
```

2

───

Now we do the same for a piecewise continuous function, representing it by a list of
functional expressions and intervals.

```
procedure integralpc(l,x);
for each v in l sum integral(first v, second v,x);
```

Example ───

```
f:={{0,{-pi,0}},{x,{0,pi}}}; integralpc(f,x);
```

```
   2
 PI
 -----
   2
```

───

Finally, the Fourier expansion itself:

```
procedure fourier(pcfunction, x,n);
begin
  scalar termcos,termsin,pcfunctioncos,pcfunctionsin;
  termcos:=integralpc(pcfunction,x)/(2*pi);
  termsin:=0;
  for j:=1:n do <<
    pcfunctioncos:=for each v in pcfunction
       collect {first(v)*cos(j*x),second v};
    pcfunctionsin:=for each v in pcfunction
       collect {first(v)*sin(j*x),second v};
    termcos:=termcos+integralpc(pcfunctioncos,x)*cos(j*x)/pi;
    termsin:=termsin+integralpc(pcfunctionsin,x)*sin(j*x)/pi>>;
  return termcos+termsin
end;
```

Example ───

```
factor sin,cos; on rat; on revpri; fourier(f,x,10);
 PI       - 2*COS(X)       - 2*COS(3*X)       - 2*COS(5*X)
 ---- + ------------- + ---------------- + ----------------
  4          PI              9*PI               25*PI
```

$$+ \frac{- 2*COS(7*X)}{49*PI} + \frac{- 2*COS(9*X)}{81*PI} + SIN(X) + \frac{- SIN(2*X)}{2}$$

$$+ \frac{SIN(3*X)}{3} + \frac{- SIN(4*X)}{4} + \frac{SIN(5*X)}{5} + \frac{- SIN(6*X)}{6}$$

$$+ \frac{SIN(7*X)}{7} + \frac{- SIN(8*X)}{8} + \frac{SIN(9*X)}{9} + \frac{- SIN(10*X)}{10}$$

3.9.3   Solving equations

The solve operator has the syntax:

$$solve(<system\ list>,<unknowns\ list>)$$

where $<system\ list>$ is a list of equations, and $<unknowns\ list>$ is a list of the unknowns the system has to be solved for. If the right-hand side of an equation is zero, it may be dropped. If there is only one equation (and one unknown), the braces may be dropped. If there are as many top-level kernels in the equations as there are unknowns, the $<unknowns\ list>$ may be dropped.

The value of the solve operator is a list of 'solutions'. If the solving of the system was successful, each 'solution' is an equation of the form

$$<identifier> = <expression>$$

In this case, the multiplicities of the solutions are stored in a global variable called multiplicities!*. Its value is a list containing the multiplicities of the most recent calculated solutions with the solve operator.

It is possible to have the solutions displayed with their multiplicities in the solution list; to do so, switch on multiplicities.

If the solve package could not find a complete solution, the 'solutions' are equations in the unknowns.

Examples ───────────────────────────────

- a general system of two linear equations in two unknowns:
  solve({a*x+b*y=f1,c*x+d*y=f2},{x,y});

$$\{\{X= - \frac{B*F2 - D*F1}{A*D - B*C},$$

```
           A*F2 - C*F1
      Y=-------------}}
           A*D - B*C
```

multiplicities!;*

{1}

* a singular system:
  *solve({x=y,3*x=3*y});*

  Unknowns: {Y,X}

  {{Y=ARBCOMPLEX(1),X=ARBCOMPLEX(1)}}

  multiplicities!;*

  {1}

* a quadratic equation:
  *solve(a*x**2+b*x+c,x);*

```
                          2
         SQRT( - 4*A*C + B ) + B
  {X= - --------------------------,
                 2*A


                      2
       SQRT( - 4*A*C + B ) - B
    X=--------------------------}
                2*A
```

  multiplicities!;*

  {1,1}

* a solvable fifth-degree equation:
  solve(x^5=1);

  Unknown: X

```
{X=1,
```

```
         - 2*SQRT(SQRT(5) - 5) - SQRT(10) - SQRT(2)
X=-----------------------------------------------,
                       4*SQRT(2)


     2*SQRT(SQRT(5) - 5) - SQRT(10) - SQRT(2)
X=---------------------------------------------,
                     4*SQRT(2)


     2*SQRT( - SQRT(5) - 5) + SQRT(10) - SQRT(2)
X=-------------------------------------------------,
                     4*SQRT(2)


         - 2*SQRT( - SQRT(5) - 5) + SQRT(10) - SQRT(2)
    X=--------------------------------------------------}
                         4*SQRT(2)
```

multiplicities!;*

{1,1,1,1,1}

- a goniometric equation:
  solve(sin x = 1/2,x);

```
         1
{X=ASIN(---) + 2*ARBINT(1)*PI,
         2



              1
     X= - ASIN(---) + 2*ARBINT(1)*PI + PI}
              2
```

multiplicities!;*

{1,1}

- a system of polynomial equations:
  *solve({x**2*y**2-1,x**3+y**3-2},{x,y});*

```
        - (SQRT(3)*I + 1)
   {{Y=--------------------,
                2
```

```
      SQRT(3)*I - 1
 X=----------------},
           2


      SQRT(3)*I - 1
{Y=--------------,
           2


     - (SQRT(3)*I + 1)
 X=--------------------},
             2


                    1/3
     - ( - SQRT(2) + 1)   *(SQRT(3)*I + 1)
{Y=-----------------------------------------,
                     2


                  2/3
 X=(( - SQRT(2) + 1)   *( - ( - SQRT(2) + 1)*SQRT(3)*I


      + ( - SQRT(2) + 1) + 2*SQRT(3)*I - 2))/2},


                1/3
     ( - SQRT(2) + 1)   *(SQRT(3)*I - 1)
{Y=---------------------------------------,
                   2


                  2/3
 X=(( - SQRT(2) + 1)   *(( - SQRT(2) + 1)*SQRT(3)*I


      + ( - SQRT(2) + 1) - 2*SQRT(3)*I - 2))/2},


                  1/3
{Y=( - SQRT(2) + 1)   ,


                2/3
 X=( - SQRT(2) + 1)   *( - ( - SQRT(2) + 1) + 2)},
```

```
                    1/3
      - (SQRT(2) + 1)    *(SQRT(3)*I + 1)
{Y=-------------------------------------,
                    2


                2/3
  X=((SQRT(2) + 1)


    *( - (SQRT(2) + 1)*SQRT(3)*I + (SQRT(2) + 1) + 2*SQRT(3)*I - 2))


    /2},


                1/3
      (SQRT(2) + 1)    *(SQRT(3)*I - 1)
{Y=----------------------------------,
                2


                2/3
  X=((SQRT(2) + 1)


    *((SQRT(2) + 1)*SQRT(3)*I + (SQRT(2) + 1) - 2*SQRT(3)*I - 2))/2}


,


                1/3
{Y=(SQRT(2) + 1)    ,


                2/3
X=(SQRT(2) + 1)    *( - (SQRT(2) + 1) + 2)},


    {Y=1,X=1}}
```

For such systems of polynomial equations REDUCE uses the Gröbner basis method.

Notice the possible introduction of arbitrary constants which, in their generality, are assumed to be complex, when the number of unknowns exceeds the number of independent equations. These arbitrary constants are numbered throughout the current REDUCE session.

The trigonometric equation (fifth example) also involves arbitrary constants in its solution, but these have to be integral. They too are numbered throughout the REDUCE session.

For transcendental equations, solve will use the known inverse functions of the built-in elementary functions. In this context we refer to section 3.6.2 where additional inverse functions have been introduced. In any case, the user should carefully check that all arguments involved are within the appropriate ranges.

3.9.4   The mkid operator

The mkid operator allows to construct a set of identifiers the name of which is composed of a fixed part (the first argument to mkid) and a varying part (the second argument to mkid). The syntax is

$$mkid(<identifier>,<expression>).$$

It can be useful in e.g. constructing a polynomial in one variable:
```
for j:=0:4 sum mkid(a,j)*x**j;
```
```
 4       3       2
X *A4 + X *A3 + X *A2 + X*A1 + A0
```
In the same order of ideas, one could think of an operator making identifiers containing several parameters. The following procedure mkidn could achieve this:
```
procedure mkidn(a,l);
begin
  scalar u,m;
  u:=a;
  m:=l;
  while m neq {} do
    <<u:=mkid(u, first m); m:=rest m>>;
  return u
end;
```

Example ——————————————————————————————————

The construction of an arbitrary polynomial in three variables of the second degree:
```
for j:=0:2 sum
for k:=0:2 sum
for l:=0:2 sum
mkidn(a,{j,k,l})*x**j*y**k*z**l;
```
```
 2 2 2           2 2           2 2         2     2
X *Y *Z *A222 + X *Y *Z*A221 + X *Y *A220 + X *Y*Z *A212
```

$$+ X^2*Y*Z*A211 + X^2*Y*A210 + X^2*Z^2*A202 + X^2*Z*A201 + X^2*A200$$

$$+ X*Y^2*Z^2*A122 + X*Y^2*Z*A121 + X*Y^2*A120 + X*Y*Z^2*A112 + X*Y*Z*A111$$

$$+ X*Y*A110 + X*Z^2*A102 + X*Z*A101 + X*A100 + Y^2*Z^2*A022 + Y^2*Z*A021$$

$$+ Y^2*A020 + Y*Z^2*A012 + Y*Z*A011 + Y*A010 + Z^2*A002 + Z*A001 + A000$$

3.9.5   The pf operator

The pf operator decomposes a rational function into its partial fractions, which are returned in a list. Its syntax is

$$pf(<expression>,<variable>).$$

Example ————————————————————————————————————

*off exp; pf(1/(1+x**4),x);*

```
     1
{--------}
    4
  X  + 1
```

This is due to the fact that there is no factorization over algebraic numbers such as sqrt 2. There is, however, a package about algebraic numbers called arnum.
load "arnum";

Now define sqrt2 to be a square root of 2:
defpoly sqrt2^2-2;

and indeed:
*sqrt2**2;*

2

Now
*pf(1/(1+x**4),x);*

```
      1
  (---*SQRT2)*(X + SQRT2)
      4
{--------------------------,
      2
   X  + SQRT2*X + 1


        1
  - (---*SQRT2)*(X - SQRT2)
        4
  --------------------------}
         2
     X  - SQRT2*X + 1
```

and to check it:
on exp; % more on the exp switch in section 3.11.3
for each v in ws sum v;

```
    1
 --------
    4
  X  + 1
```

3.9.6 The length operator

We already discussed this operator in connection with lists, arrays and matrices. More generally it also enables the user to find the length of an expression, it being the number of additive top-level terms in the expanded representation of this expression.

Examples ───────────────────────────────

length x;
1
length(x + sin x);
2
*length (y**2 + tan x)**3;*
8
*length (x*y);*
1
length(x/z);
2

```
length(x*y/z);
2
```

In view of the last three examples, we advise the user to handle the length operator acting on expressions with great caution.

3.9.7 The coeff operator

The coeff operator splits up an algebraic expression in terms of increasing powers of a given kernel and returns a list of the coefficients of these terms. Its syntax is:

$$\mathtt{coeff(<\mathit{expression}>,<\mathit{kernel}>)}$$

Normally it is assumed that $<expression>$ is polynomial in $<kernel>$, otherwise an error message results.

Examples ——

```
ex:=(x+1/y)**5;
         5 5        4 4        3 3         2 2
         X *Y  + 5*X *Y  + 10*X *Y  + 10*X *Y  + 5*X*Y + 1
EX  := ---------------------------------------------------
                                  5
                                  Y

coeff(ex,x);
    1     5    10    10    5
  {----,----,----,----,---,1}
    5     4    3     2    Y
    Y     Y    Y     Y

coeff(ex,y);
          5 5        4 4        3 3         2 2
          X *Y  + 5*X *Y  + 10*X *Y  + 10*X *Y  + 5*X*Y + 1
  *****-------------------------------------------------------- invalid as
                                   5
                                   Y


POLYNOMIAL
```

However, if $<expression>$ is rational in $<kernel>$ and ratarg is switched on, the denominators will not be checked for dependence on $<kernel>$ and the powers of $<kernel>$ in the denominators are simply included in the coefficients.

Example ————————————————————————————————

```
on ratarg; coeff(ex,y);
            2         3       4     5
  1    5*X    10*X    10*X    5*X    X
{----,-----,-------,-------,-----,----}
  5     5      5       5       5     5
  Y     Y      Y       Y       Y     Y
```

The same happens when $<kernel>$ appears within a functional form. The coeff operator does not check for this disguised $<kernel>$ and includes it in the coefficient.

Example ————————————————————————————————

```
coeff(x*sin(x),x);
{0,SIN(X)}
```

Each time the operator coeff is evaluated, the variables hipow!* and lowpow!* are set to the highest, resp. the lowest degree of $<kernel>$ encountered in $<expression>$. If the user is interested in one particular coefficient, then instead of manipulating the list procured by the operator coeff, he can directly ask for that coefficient by

$$\text{coeffn}(<expression>,<kernel>,<integer>)$$

where $<integer>$ represents the power of $<kernel>$ in the term the coefficient is asked for. This is the element at location $<integer>+1$ in the list returned by coeff($<expression>$,$<kernel>$).

Example ————————————————————————————————

```
coeffn(ex,x,1);
  5
----
  4
 Y
```

3.10   Polynomials

REDUCE is particularly suited to handling polynomials and rational functions. In the previous sections, we already commented on several operators that are also applicable in this context. In this section, we focus on operators and switches that are typical for polynomial and rational function manipulation.

3.10.1   Factorization of polynomials and rational functions

A key problem for polynomial manipulation is factorization. In REDUCE, factorization of univariate and multivariate polynomials with coefficients in the ring of integers or even in other domains (such as modular arithmetic) is possible.

A direct factorization of a polynomial is obtained by the procedure `factorize` with syntax:

$$\texttt{factorize}(<expression>)$$

where $<expression>$ should be a polynomial, otherwise an error message results. The value of this operator is a list containing all factors with integer coefficients. The order of the factors is system-dependent and one should not rely on it.

If, however, there is an overall numerical factor, it appears as the first in the list. If the user wants this numerical factor to be factored into prime factors, the switch `ifactor` should be on.

Examples ──

```
on ifactor; factorize(1024*(x**3-y**3));
                       2        2
{2,2,2,2,2,2,2,2,2,2,X  + X*Y + Y ,X - Y}
rat1:=(x**2-y**2)/(u**3+v**3);
            2   2
           X - Y
RAT1  :=  ---------
           3   3
           U + V
factorize rat1;
          2   2
         X - Y
*****  ----------- invalid as POLYNOMIAL
          3   3
         U + V
```

──

As the last expression in the example was rational, no factorization was hoped for with this operator. An alternative—which indeed works for rational expressions—is the switch `factor`. When it is on, the numerator and denominator will be searched for factors and the result will be the original expression, in factored form.

```
on factor; rat1;
       (X + Y)*(X - Y)
     -------------------------
        2          2
     (U  - U*V + V )*(U + V)
```

In evaluating a rational function, REDUCE cancels common factors shared by the numerator and denominator. This cancellation process is reinforced by turning the switch gcd on, in which case the greatest common divisor of the numerator and denominator is calculated. There is a second, faster algorithm for computing greatest common divisors, activated by the switch ezgcd.

Examples ───

When the denominator divides the numerator, the division is carried out automatically:

```
(a**7+3*a**5*b**2+2*a**4*b**4+a**3*b**6+6*a**2*b**6+2*b**10)
/(a**4+3*a**2*b**2+b**6);

 3       4
A   + 2*B
```

If gcd is off, common factors may not be cancelled:

```
a:=(x-1)*(u**4+u**2+1);

        4       4     2         2
A  := U *X  -  U  + U *X  -  U  + X  -  1

b:=(y**2+1)*(u**4+2*u**3+u**2-1);

        4  2     4       3  2       3     2  2     2     2
B  := U *Y  +  U  + 2*U *Y  + 2*U  + U *Y  + U  -  Y  -  1

a/b;

          4       4     2         2
         U *X  -  U  + U *X  -  U  + X  -  1
--------------------------------------------------------

  4  2     4       3  2       3     2  2     2     2
 U *Y  +  U  + 2*U *Y  + 2*U  + U *Y  + U  -  Y  -  1
```

When it is switched on:

```
on gcd; a/b;

   2       2
  U *X  -  U  - U*X + U + X - 1
------------------------------------

  2  2     2       2         2
 U *Y  +  U  + U*Y  + U  -  Y  -  1
```

The gcd switch is off by default, because it can be time-consuming and because rational expressions may become apparently *more* complicated after cancellation of common factors. Let us give a few more illustrations:

Examples ───

```
off gcd;  z:=(a**10-b**10)/(a**3-b**3);
```

```
       10    10
      A    - B
Z  := -----------
       3    3
      A  - B
```

on factor; on time; z;

```
   4    3      2 2     3    4    4    3      2 2     3    4
((A  + A *B + A *B  + A*B  + B )*(A  - A *B + A *B  - A*B  + B )
```

```
            2        2
 *(A + B))/(A  + A*B + B )
```

Time: 901 ms   plus GC time: 170 ms

off factor,time; on gcd; z;

```
   9    8      7 2    6 3    5 4    4 5    3 6    2 7      8
(A  + A *B + A *B  + A *B  + A *B  + A *B  + A *B  + A *B  + A*B
```

```
   9    2        2
 + B )/(A  + A*B + B )
```

3.10.2   The functions gcd, lcm and mcd.

To compute directly the greatest common divisor of two polynomials, use

$$\text{gcd}(<expression~1>,<expression~2>)$$

where both expressions must be polynomials.

Whenever an expression contains the sum of two rational expressions REDUCE will add them by putting them over a common denominator which is the least common multiple of both denominators.

Example ─────────────────────────────────────

u:=a/((x+1)(x+2)*(x+3));*

```
                A
U  := ----------------------
       3      2
      X  + 6*X  + 11*X + 6
```

v:= b/((x+1)(x+2)*(x+4));*

```
                B
V  := ----------------------
       3      2
      X  + 7*X  + 14*X + 8
```

```
w:=u+v;
        A*X + 4*A + B*X + 3*B
W  := --------------------------------
       4      3       2
       X  + 10*X  + 35*X  + 50*X + 24
```

This occurs as long as the switch lcm is on, as is normally the case. If lcm is switched off, REDUCE still produces a rational expression over a common denominator, but this is now simply the product of both denominators.

Example ———————————————————————————

```
off lcm; u+v;
     3       2                       3        2
  A*X  + 7*A*X  + 14*A*X + 8*A + B*X  + 6*B*X  + 11*B*X + 6*B
--------------------------------------------------------------
      6       5       4        3        2
      X  + 13*X  + 67*X  + 175*X  + 244*X  + 172*X + 48
```

If it is not desirable to have denominators combined, the user can switch off the switch mcd, 'make common denominator', which controls this process; mcd is normally on. With mcd off, denominators are treated as negative powers—this explains why only very few combinations occur.

Examples ———————————————————————————

```
off mcd; uu:=a/((x+1)*(x+2)*(x+3));
          3     2           -1
UU := (X  + 6*X  + 11*X + 6)  *A
vv:=b/((x+1)*(x+2)*(x+4));
          3     2            -1
VV := (X  + 7*X  + 14*X + 8)  *B
uu+vv;
    3     2           -1
(X  + 7*X  + 14*X + 8)  *B +

    3     2           -1
(X  + 6*X  + 11*X + 6)  *A
```

Switching off mcd should be done with the necessary caution: the canonical form of the output expressions is not longer guaranteed, which means that rational expressions equivalent to zero no longer simplify to zero.

The least common multiple of two polynomials can also be computed directly by

$$\text{lcm}(<expression\ 1>,<expression\ 2>)$$

where both expressions have to be polynomials.

3.10.3   resultant

Given two univariate polynomials p and q, of degree n and m respectively:

$$p(x) = a_0 + a_1 x + \ldots + a_n x^n, \quad q(x) = b_0 + b_1 x + \ldots + b_m x^m,$$

their resultant $R(p,q)$ is, by definition, the determinant of the following $(m+n) \times (m+n)$-matrix:

$$
\begin{bmatrix}
a_n & a_{n-1} & \cdots & a_{n-m+1} & a_{n-m} & a_{n-m-1} & \cdots & a_1 & a_0 & 0 & \cdots & 0 \\
0 & a_n & \cdots & a_{n-m+2} & a_{n-m+1} & a_{n-m} & \cdots & a_2 & a_1 & a_0 & \cdots & 0 \\
\vdots & \vdots & \ddots & \vdots & \vdots & \vdots & & \vdots & \vdots & \vdots & \ddots & \vdots \\
0 & 0 & \cdots & a_n & a_{n-1} & a_{n-2} & \cdots & a_m & a_{m-1} & a_{m-2} & \cdots & a_0 \\
b_m & b_{m-1} & \cdots & b_1 & b_0 & 0 & \cdots & 0 & 0 & 0 & \cdots & 0 \\
0 & b_m & \cdots & b_2 & b_1 & b_0 & \cdots & 0 & 0 & 0 & \cdots & 0 \\
\vdots & \vdots & \ddots & \vdots & \vdots & \vdots & & \vdots & \vdots & \vdots & \ddots & \vdots \\
0 & 0 & \cdots & b_m & b_{m-1} & b_{m-2} & \cdots & 0 & 0 & 0 & \cdots & 0 \\
\vdots & \vdots & \ddots & \vdots & \vdots & \vdots & & \vdots & \vdots & \vdots & & \vdots \\
0 & 0 & \cdots & 0 & 0 & 0 & \cdots & b_m & b_{m-1} & b_{m-2} & \cdots & b_0
\end{bmatrix}
$$

Some of its properties are:

- $R(p,q) = (-1)^{nm} R(q,p)$

- $R(a_0,q) = a_0^m$

- $R(a_0,b_0) = 1$

- $R(p,q) = 0$ iff p and q have a common polynomial factor

- if α_i, $i = 1, \ldots n$ are the zeroes of p,

$$R(p,q) = a_n^m q(\alpha_1) \cdots q(\alpha_n),$$

and, similarly, if the β_j, $j = 1, \ldots, m$ are the zeroes of q,

$$R(p,q) = (-1)^{mn} b_m^n p(\beta_1) \cdots p(\beta_m);$$

both formulas imply that

$$R(p,q) = a_n^m b_m^n \prod_{i=1}^{n} \prod_{j=1}^{m} (\alpha_i - \beta_j).$$

The command

$$\text{resultant}(<expression\ 1>,<expression\ 2>,<identifier>)$$

computes the resultant of the polynomials $<expression\ 1>$ and $<expression\ 2>$ with respect to the variable $<identifier>$.

3.10.4 remainder

The remainder function acts on two (univariate) polynomials and returns the remainder of their quotient. The syntax is

$$\text{remainder}(<expression\ 1>,<expression\ 2>)$$

where both expressions are polynomials.

Example ——————————————————————————————————

```
remainder(x**3+1,x+1);
```

0

```
remainder(x**3+2*x**2+3*x+4,x+a);
```

```
   3       2
X  + 2*X  + 3*X + 4
```

This last result is not really what we expected. This is due to the fact that during the calculation of the remainder the internal ordering is respected, which makes the second polynomial to be interpreted as a polynomial in the variable a. Changing this ordering by

```
korder x;
```

yields:

```
remainder(x**3+2*x**2+3*x+4,x+a);
```

```
     3       2
  - A  + 2*A  - 3*A + 4
```

as desired.

———

3.10.5 decompose

The operator decompose acts upon a multivariate polynomial and returns a list containing a (not unique) tree structure of that polynomial.

Example ——————————————————————————————————

```
w:=x-2*y; v:=3*w**2-7*w+1; ex:=v**3-v+5; decompose ex;
```

```
   3       2
{A  + 3*A  + 2*A + 5,
```

```
      2
A=3*B   - 7*B,

B=X - 2*Y}
```

3.10.6   interpol

The operator interpol has the syntax

$$interpol(<list\ 1>,<variable>,<list\ 2>)$$

where $<list\ 1>$ (a list of values) and $<list\ 2>$ (a list of points) are lists of equal length, say l.

The result is a polynomial in the variable $<variable>$ of degree less than or equal to $(l-1)$, with the property that the value of that polynomial in a 'point' of $<list\ 2>$ is the corresponding 'value' of $<list\ 1>$.

Example ──

```
points:={a,b,c}; values:={u,v,w}; operator p;
let p(~x) = interpol(values,x,points);
p(y);
   2          2         2        2          2         2          2
(A *B*W  - A *C*V  + A *V*Y - A *W*Y - A*B *W  + A*C *V - A*V*Y


        2    2          2       2          2          2          2
  + A*W*Y  + B *C*U - B *U*Y + B *W*Y - B*C *U + B*U*Y   - B*W*Y


     2          2              2      2   2      2          2        2
  + C *U*Y - C *V*Y - C*U*Y   + C*V*Y )/(A *B - A *C - A*B  + A*C


     2      2
  + B *C - B*C )
```

and to check this result

```
for each q in points collect p(q);
{U,V,W}
```

3.10.7   deg

The deg function, with syntax

$$deg(<expression>,<kernel>)$$

returns the highest degree in the kernel $<kernel>$ that occurs in the polynomial $<expression>$.

3.10.8   den and num

The functions den and num, with syntax

$$\mathrm{den}(<expression>)$$

and

$$\mathrm{num}(<expression>)$$

return the denominator, resp. the numerator of the expression.

3.10.9   lcof, lterm and reduct

The function lcof with syntax

$$\mathrm{lcof}(<expression>,<kernel>)$$

returns the leading coefficient of the kernel $<kernel>$, in the $<expression>$, which must be a polynomial. If it does not occur in $<expression>$, $<expression>$ itself is returned.

The function lterm with syntax

$$\mathrm{lterm}(<expression>,<kernel>)$$

returns the leading term with respect to the kernel $<kernel>$ in the polynomial $<expression>$. If the kernel $<kernel>$ does not occur in $<expression>$, zero is returned.

As is naturally expected, the leading coefficient of a variable in a polynomial is also the leading coefficient of the leading term with respect to that variable in the polynomial. But if the variable does not appear in the polynomial, this is no longer the case following the rules explained above. These exceptional cases must also be taken care of when using the following function, reduct.

This function acts on a polynomial, has the syntax

$$\mathrm{reduct}(<expression>,<kernel>)$$

and returns the part of $<expression>$ that remains after removing the leading term with respect to the kernel $<kernel>$. If the polynomial is a number, zero or not, that number is returned. If $<kernel>$ does not occur in $<expression>$, zero is returned. Except for these special cases, one has the identity:

$$<expression> = \mathrm{lterm}(<expression>,<kernel>)$$
$$+\mathrm{reduct}(<expression>,<kernel>)$$

3.10.10   mainvar

The `mainvar` function acts on a polynomial, with the syntax

$$\text{mainvar}(<expression>)$$

It returns the variable that is ordered first in the internal ordering and occurs in the polynomial $<expression>$. If $<expression>$ reduces to a number in the coefficient domain, zero is returned.

3.10.11   Rational coefficients

Up to now, we have tacitly assumed that all coefficients in the polynomials are integers. This implies that a polynomial such as 1/2*x is regarded as a rational expression on which many functions (such as `lterm` and `lcof`) cannot be applied—if one tries, an error message results.

In order to admit rationals as coefficients in a polynomial (which then is no longer regarded as a rational expression), it suffices to switch on `rational`.

Examples ───

```
p:=(1/2*x*y**2+3/5*u**3*v)**2;

            6  2        3     2        2 4
        36*U *V   + 60*U *V*X*Y   + 25*X *Y
P  := ------------------------------------------
                        100

lcof(p,x);

            6  2        3     2        2 4
        36*U *V   + 60*U *V*X*Y   + 25*X *Y
***** ------------------------------------------ invalid as polynomial
                        100

on rational; p;

 9      6 2    5  3        2     25   2 4
----*(U *V   + ---*U *V*X*Y   + ----*X *Y )
 25             3                36

lcof(p,x);

  1   4
 ---*Y
  4
```

3.10.12 Floating-point coefficients

We already discussed the switch rounded in section 3.1 on numbers. When rounded is on, any rational number is converted into a fixed-precision decimal number—the precision is system-dependent—by putting the denominator equal to the integer 1 and adjusting the numerator.

Examples ───

```
x:=2*a/3;

      2*A
X := -----
       3

den x;

3

num x;

2*A

lcof(x,a);

      2*A
***** ----- invalid as polynomial
       3

on rounded; x;

0.666666666667*A

den x;

1

num x;

0.666666666667*A

lcof(x,a);

0.666666666667
```

The same considerations about the conversion of rational numbers to decimal numbers when rounded is on, hold for algebraic and transcendental coefficients entered as the functional values of certain mathematical functions such as the square-root function, the sine function, etc.

Example ──

```
on numval; x:=a/sqrt(3);

X := 0.57735026919*A

den x;

1
```

num x;

0.57735026919*A

lcof(x,a);

0.57735026919

3.10.13   Complex coefficients

We already encountered complex numbers in section 3.1.5. The reserved variable I stands for a square root of (-1), and all algebraic operations go through on numbers of the form a+b*i, where a and b are integer, rational, algebraic or transcendental.

Examples ───────────────────────────────────

*(a+b*i)*(a-b*i);*

$$A^2 + B^2$$

Do not expect the resulting expression to be factorized over the complex numbers:
factorize ws;

$$\{A^2 + B^2\}$$

If factorization over the complex numbers is required, switch on complex:
*on complex; factorize(a**2+b**2);*

{A - I*B,A + I*B}

The basic functions conj, repart and impart (Cf. also section 3.6.2) are also applicable on non-numeric complex numbers; the real and imaginary parts are interpreted as being complex too.

Example ────────────────────────────────────

*z1:=a+b*i; z2:=conj(1/z1);*

Z2 := (IMPART(A)*I - IMPART(B) + REPART(A) + REPART(B)*I)/(IMPART(A)

$$+ 2*IMPART(A)*REPART(B) + IMPART(B)^2$$

$$- 2*IMPART(B)*REPART(A) + REPART(A)^2 + REPART(B)^2)$$

When used as coefficients in polynomials, and in order to allow the above discussed polynomial operations under default switch settings, those complex numbers must have the form a+b*i where a and b are integers.

If the real and imaginary parts of the complex coefficients are rational, rational has to be switched on. If they are algebraic or transcendental, rounded has to be switched on.

Example ───

```
p1:=z1*x;
P1 := X*(A + B*I)
lcof(p1,x);
A + B*I
p2:=sub(a=3/4,b=5/12,p1);
        5              9
P2 := ----*X*(I + ---)
        12             5
on rational; lcof(p2,x);
  5          9
----*(I + ---)
  12         5
p3:=sub(a=cos(pi/8),b=sin(pi/8),p1);
              1                    1
P3 := X*(SIN(---*PI)*I + COS(---*PI))
              8                    8
on rounded; lcof(p3,x);
*** Domain mode RATIONAL changed to ROUNDED

0.382683432365*I + 0.923879532511
```
───

If the coefficients of a polynomial contain the reserved identifier I in the denominator, this polynomial is considered as a rational expression (as is naturally the case when a clear variable appears in the denominator).

If rationalize is switched on, complex denominators are changed to real ones by multiplying by the complex conjugated denominator. If, in addition, rational is switched on, a 'polynomial' with rational complex coefficients (with real denominator) really is considered as a polynomial, and the polynomial manipulations go through.

Examples ──

```
1/z1;
```

```
     1
 ---------
  A + B*I
```
on rationalize; 1/z1;
```
  A - B*I
 ---------
   2    2
  A  + B
```

3.11 Structure.

Up to a certain degree, the user can control the output format of the values of the expressions, and even the internal operations. This is done by setting or clearing various switches, using the on and off commands, or by declarations using some prefix operators such as order and korder.

Of course, an on command only has an effect if the current status of the switch was off, and vice-versa. The default status of all switches is therefore important.

An important notion in the structural context is that of a kernel. To introduce it, we recall two examples: in the section on operators we encountered the functional value sin(x+y), which evaluated to itself. In the section on differentiation, we encountered df(y,x) for the derivative of an unspecified function y dependent on the variable x. Both expressions are values of an operator (sin or df) at given arguments, which cannot be simplified any further. They are called kernels and are essentially treated as if they were clear variables; clear variables themselves are also considered to be kernels.

Kernels always arise from the evaluation process of expressions when terms containing the original operator remain when all possible simplifications have been carried out. This is not the case with e.g. the arithmetic operators because these operators do not appear anymore in the canonical form where the simplification rules lead to. So a*b is not a kernel, whereas sqrt(a*b) is.

We now describe the effects of some of REDUCE's algebraic-mode switches.

3.11.1 output

Switching off the output switch has a dramatic effect: all output of any evaluation is suppressed. This has only restricted usefulness, e.g. when loading files. Needless to say, the default status of output is on.

Example ——

off output;
ex:=x(a*x**2+c*y**5*z**2)+x*(b*x*y*(z-u)**3+d*u)/(2*v);*

(no output)

on output; ex;

$$\begin{array}{rcl}
& 2 & 3 & 2 & 2 & 3 \\
(X*(2*A*V*X & - B*U & *X*Y + 3*B*U & *X*Y*Z - 3*B*U*X*Y*Z & + B*X*Y*Z & + 2*C
\end{array}$$

$$\begin{array}{c}
5 \ 2 \\
*V*Y \ *Z \ + D*U))/(2*V)
\end{array}$$

In the sequel, we will refer to this output of ex's value as the 'original' one.

3.11.2 pri

Transforming the output format of large expressions may consume a significant amount of computer time. Speeding up the printing of the output is possible by switching off pri: the output then takes a fixed format that reflects the internal structure of the expressions, and on which output switch settings have no influence. pri is on by default.

With pri off, our expression ex is printed as:

$$\begin{array}{cccccccc}
3 & 2 \ 3 & 2 \ 2 & 2 & 2 & 3 & 2 \\
(2*X \ *V*A + (& - Y*X \ *U & + 3*Z*Y*X \ *U & - 3*Z \ *Y*X \ *U & + Z \ *Y*X \ )*B + 2*Z
\end{array}$$

$$\begin{array}{c}
2 \ 5 \\
*Y \ *X*V*C + X*U*D) \ / \ 2*V
\end{array}$$

3.11.3 exp

In the original output of ex's value, we notice that the factor (z-u)**3 has been expanded. Expansion of expressions is controlled by the switch exp. When it is turned off, no expansion of powers or products of expressions occurs, and two equivalent expressions may no longer simplify to the same value. However, a zero expression is still simplified to zero. exp is usually on.

If we print the value of ex with exp off, we get

$$\begin{array}{ccc}
3 & 2 & 5 \ 2 \\
(((U - Z) \ *B*X*Y - D*U) - 2*(A*X & + C*Y \ *Z \ )*V)*X
\end{array}$$

$$- \ \frac{\rule{9cm}{0.4pt}}{2*V}$$

3.11.4 order

In the original output of ex's value, the first term of the second factor of the numerator has a as its first factor. The second up to the fifth term start with the factor b, etc. This is no coincidence, but reflects the internal ordering REDUCE imposes on identifiers appearing in expressions. This ordering is system-dependent, but in most cases alphabetical. As it can affect the intelligibility of the output, it is quite handy to have a set of commands that can influence this ordering.

A first ordering declaration uses the command order with syntax:

$$\text{order } <kernel\ 1>,<kernel\ 2>,\ldots$$

which will order $<kernel\ 1>$ ahead of $<kernel\ 2>$ etc.

A second order command will rearrange the order of the kernels involved, but the newly mentioned kernels will automatically rank lower than the previously mentioned. The default ordering for output is reset by the command order nil. It should be emphasized that the order command in no way affects the internal ordering of the kernels used during the calculations. It only influences the produced output.

Examples ───────────────────────────────────

```
order x,y,z,v; ex;
        2              3            2                    2           3
(X*(2*X *V*A + X*Y*Z *B - 3*X*Y*Z *B*U + 3*X*Y*Z*B*U  - X*Y*B*U  + 2*

      5  2
      Y *Z *V*C + D*U))/(2*V)
order y,x; ex; order nil;
```
(The order is now z, v, y, x.)
```
        3            2   5           2                          2           2
(X*(Z *Y*X*B + 2*Z *V*Y *C - 3*Z *Y*X*B*U + 3*Z*Y*X*B*U  + 2*V*X *A -

          3
      Y*X*B*U  + D*U))/(2*V)
```
───

Notice that in an expression or a sub-expression enclosed by parentheses, the terms are arranged according to decreasing powers of the first kernel in the user-given hierarchy.

3.11.5   korder

If the internal ordering of the kernels has to be changed in order to influence the calculations (e.g. to save computer time), one should use the command korder, with syntax

$$\text{korder } <kernel\ 1>,<kernel\ 2>,\ldots$$

The internal order is reset to the system default by the command korder nil.

3.11.6 factor

In the original output of ex's value, several terms contain the same combination of powers of the variable x. It might be interesting to gather all these terms and to factor out each combination of powers. This is performed by giving to the selected variables the factor status, using the command

factor <kernel 1>,<kernel 2>,...

which will factor out all combinations of powers of the mentioned kernels.
To remove a kernel from this 'factoring list', use the command

remfac <kernel>

Notice that the factor command does not factorize, but merely affects the output. Do not confuse it with the factor switch (as in 'on factor;') or with the factorize operator.
Further factor declarations add the newly mentioned kernels to the already existing 'factoring list'. The more kernels are on this 'factoring list', the fewer grouping of terms occurs.

Examples ————————————————————————————————————

```
factor x; ex;
     3        2           3       2         2   3              5   2
(2*X *A*V + X *B*Y*( - U   + 3*U *Z - 3*U*Z   + Z ) + X*(2*C*V*Y *Z   + D

    *U))/(2*V)
factor z; ex; remfac x,z;
     3        2  3          2  2          2       2       2   3
(2*X *A*V + X *Z *B*Y - 3*X *Z *B*U*Y + 3*X *Z*B*U *Y - X *B*U *Y + 2*

    2     5
X*Z *C*V*Y   + X*D*U)/(2*V)
```

——

3.11.7 allfac

In the previous examples, it could be seen that overall simple multiplicative factors, be it numbers, unbound variables or functional values, possibly raised to an integer power, in the whole expression or in sub-expressions enclosed by parentheses, are factored outside the parentheses. This was for instance the case with the factor x in the expression ex. This process is governed by the switch allfac, which is on by default. When it is desirable to inspect the output fully expanded, this monomial factorization can be suppressed by switching allfac off.

Example ──

off allfac; ex; on allfac;

```
         3       3 2         2 2              2   2       2   3
(2*A*V*X  - B*U *X *Y + 3*B*U *X *Y*Z - 3*B*U*X *Y*Z  + B*X *Y*Z  + 2*


       5 2
 C*V*X*Y *Z  + D*U*X)/(2*V)
```

──

3.11.8 div

When the div switch is on, simple numerical and monomial factors in the denominator of an expression are divided in the numerator, so that rational fractions and negative powers will emerge. Normally, div is off.

Example ──

on div; ex; off div;

```
       2   1      3  -1       3     2  -1       3        -1     2
X*(A*X   - ---*B*U *V  *X*Y + ---*B*U *V  *X*Y*Z - ---*B*U*V   *X*Y*Z +
           2                  2                    2


     1      -1    3       5 2   1        -1
    ---*B*V   *X*Y*Z  + C*Y *Z  + ---*D*U*V   )
     2                            2
```

──

3.11.9 list

The reading and inspection of a result is sometimes made easier when each term of a sum is printed on a separate line. This is achieved by switching on list, which is normally off.

Example ──

on list; ex; off list;

```
          2
(X*(2*A*V*X


       3
    -B*U *X*Y
```

```
           2
  +3*B*U *X*Y*Z


              2
  -3*B*U*X*Y*Z


            3
  +B*X*Y*Z


           5  2
  +2*C*V*Y *Z


  +D*U))/(2*V)
```

3.11.10 rat

It is sometimes convenient, e.g. in polynomials with rational coefficients or in truncated power series, not to have one common denominator, but to distribute them over all subexpressions.

This is accomplished by switching on rat (it is off by default), in combination with the factor command, in which case the overall denominator of the expression will now divide each factored subexpression.

If no unbound variable has factor status, the rat switch has no effect.

Example ——————————————————————————————————————

```
factor x; on rat; ex; off rat; remfac x;
                    3      2      2    3             5  2
   3   ' 2      - U  + 3*U *Z - 3*U*Z  + Z      2*C*V*Y *Z  + D*U
  X *A + X *B*Y*--------------------------- + X*-------------------
                          2*V                           2*V
```

The user should experiment with the interaction of the commands order and factor, the switches exp, mcd and gcd which affect computations and the output switches allfac, rat and div. A combination of all these provides a wide variety of output forms.

3.11.11 ratpri

The ratpri switch is normally on. It causes a rational expression to be printed in the natural way, i.e. the numerator and the denominator on separate lines with the fraction line in between, but only if both, the numerator and denominator can be printed on one line. If this is not possible, or if ratpri is off, the expression is printed with the quotient slash.

Example ────────────────────────────────────

```
ex2:=(a+b)/(a-b);
        A + B
EX2  := -------
        A - B
off ratpri; ex2; on ratpri;
(A + B)/(A - B)
```

───

3.11.12   revpri

We already noticed that in an expression, or in a sub-expression enclosed by brackets, the terms are ordered in decreasing powers of the kernel with the highest rank in the current ordering. It might be interesting, e.g. in a truncated power series expansion, to order with respect to increasing powers. This is achieved by switching on revpri, whose default status is off.

Example ────────────────────────────────────

We define a procedure to compute the Taylor series expansion of a function up to a certain degree:

```
procedure taylor(f,a,n);
begin
    scalar g,term;
    g:=f;
    term:=sub(x=a,g);
    for j:=1:n do
        <<
            g:=df(g,x)/j;
            term:=term+sub(x=a,g)*(x-a)**j
        >>;
    return term
end;

TAYLOR
```

Now apply it:

```
taylor(sin x, 0, 5);
      4      2
 X*(X  - 20*X  + 120)
 --------------------
         120
```

Notice the effect of the rat and revpri switches:

```
factor x; on rat; on revpri; ws;
off revpri; off rat; remfac x;
         3       5
       - X       X
X + ------- + -----
         6      120
```

3.11.13   nero

The switch nero, when activated (the default status is off), suppresses the printing of zero expressions.

3.12   Internal representation

3.12.1   The share command

Normally, the algebraic value of a variable is not stored in its LISP value, but as one of its properties, so that it can be difficult to manipulate in a LISP procedure. The share command ensures that the value is stored in the LISP value.

Example ───

```
l:=(a-b)**2;
      2               2
L := A  - 2*A*B + B
lisp l;
***** 'L' is an unbound ID
share l; l;
      2               2
L := A  - 2*A*B + B
lisp l;
(!*SQ ((((A . 2) . 1) ((A . 1) ((B . 1) . -2)) ((B . 2) . 1)) . 1) T)
```

3.12.2   The part operator

The starting point for the explanation of this operator is the internal prefix form of the algebraic expression under consideration, as it is printed or as it would have been printed taking all relevant switch settings into account. It thus is assumed that the switch pri is on, since no prefix form of an expression is constituted (but only a representation tree) if pri is off.

As a prefix form of an expression is a list, the operator part with syntax

$$part(<expression>,<integer>)$$

takes the element of rank $<integer>$ out of this list, while the operator `arglength` with syntax

$$arglength(<expression>)$$

returns one more than the length of this list. Hence the command
part(ex,0);
always returns the top-level operator in the expression.

Examples ——

ex:=(x+1/y)**5; part(ex,1);

```
 5  5      4  4          3  3          2  2
X *Y  + 5*X *Y  + 10*X *Y  + 10*X *Y  + 5*X*Y + 1
```

part(ex,0);

QUOTIENT

part(ex,-1);

```
 5
Y
```

the last result being in accordance with the working of the operator `length` on lists.

——

As already mentioned, the working of the operator `part` becomes transparent when looking down at the prefix form of the algebraic expression, which we can obtain by applying the LISP function `reval` (part of the REDUCE package) to 'ex:
lisp reval 'ex;

(quotient (plus (times (expt X 5) (expt Y 5)) (times 5 (expt X 4)
(expt Y 4)) (times 10 (expt X 3) (expt Y 3)) (times 10 (expt X 2)
(expt Y 2)) (times 5 X Y) 1) (expt Y 5))

This also explains the value of
part(ex,1,1,1,1);

X

which in fact stands for
part(part(part(part(ex,1),1),1),1)

The operator part can also be used to substitute another expression for a particular part of a given expression. The syntax then reads:

$$part(<expression\ 1>,<integer>) \ := \ <expression\ 2>$$

and the result is $<expression\ 1>$ after the substitution.

Example ————————————————————————————————————

```
part(ex,2):=sin z;
   5  5      4  4        3  3        2  2
   X *Y  + 5*X *Y  + 10*X *Y  + 10*X *Y  + 5*X*Y + 1
  ----------------------------------------------------
                          SIN(Z)
```

3.13   Files.

It often happens during interactive REDUCE sessions that complicated commands
or simplification rules or extended procedure definitions are to be re-typed, usually
with minor changes, due to some programming errors. It is then advantageous to
write these REDUCE instructions in a file by means of a text editor. This is even
more the case when developing yourself a set of REDUCE programs for a particular
mathematical topic.

Be alert to end such file by

$$<terminator> \text{ end } <terminator>$$

—the reason for this is given below.

In the course of an interactive REDUCE session, this file can be read in at any
moment. The appropriate command for loading the file $<user\ file>$ (local file-naming
conventions of course apply) is:

$$\text{in } <user\ file>$$

or

$$\text{in } "<user\ file>"$$

Notice the use of the double-quotes, which is system-dependent.

If the in command is terminated by a semicolon, all instructions, together with the
answers (i.e. the values of the subsequent commands), will appear on the screen. If
the terminator is a dollar sign, the commands will not be shown, but the results will
be, at least if the instructions ended by a semicolon.

Mostly, a REDUCE file only contains procedure definitions, simplification rules, etc.
which have no value assigned to them. In this case it is desirable not to have this file
printed out each time it is loaded. The printing of a file can be governed by the echo
switch.

If the command off echo is included in a file, the rest of that file from that point on
will not be printed out, even if semicolon-ended commands were used. Conversely, if
on echo appears in a file, everything from that point on is printed, even if dollar signs
are used as terminators. So, if the statement off echo is inserted at the beginning of

a file, in combination with on echo at the end of it, the file on its whole will not be printed out when loading it and the printing options will be restored to their default status afterwards.

When a file is read in, it is processed until an end is encountered which is not matching a preceding begin. So ending up a file with e.g. ;end; is a precaution to closing the reading of the file, no matter which programming error (such as the omission of an end) was committed. In certain implementations it is even compulsory.

There are some extra commands which influence the reading of a file in REDUCE. The processing of a file is momentarily interrupted if the command

pause

is encountered. It allows the user to perform some interactive computations, probably in connection with the results emerging from the file's commands. It may even be inserted in a procedure definition if input from the user is needed there. By the command

cont

('continue') the processing of the file is resumed. A cont command is only effective after a preceding pause command. Otherwise it leads to an error message.

When the demo switch is on, REDUCE pauses after the execution of each command. This is particularly useful for a REDUCE demonstration, because it offers the opportunity for verbal explanation. This command can be included in the file, but it can also typed in directly before the file is loaded.

Written comments can be inserted at any moment in a file. REDUCE considers the string of characters in between the word comment and the next terminator as commentary and does not act on it. Also, the whole line following the percent sign % is considered as commentary.

Let us mention a difference between an interactive REDUCE session and the reading of a file, concerning operator declaration. Perhaps you have noticed that if you type in a statement of the form:

f(a,b,c);

REDUCE replies with the question:

Declare f operator?

If the same statement appears in a file, this question will not be asked explicitly and, an affirmative answer being assumed, the message f declared operator

is printed out. This is due to the fact that during input from a batch file loaded by an in command, the switch int, which is on by default, is automatically put off. When int is on, REDUCE will pause and query the user if it encounters anomalies such as an undeclared operator. Under the same circumstances, when int is off, REDUCE will proceed. One can even suppress possible error messages by switching off msg.

If the user wishes to save certain expression assignments needed later on, before leaving an interactive REDUCE session, he can write them to a file. The output is redirected from the screen to the file named *<user file>* by the command

$$out \ \ <user \ file>$$

or

$$out \ \ "<user \ file>"$$

This causes the file to be opened and all subsequent output to be written on it until either the file is shut, another file is opened, or the output is redirected to the screen by the command

$$out \ t$$

where t stands for 'terminal'.

Notice that statements without value such as procedure definitions, let rules, declarations etc. produce no output and are not written on the open file.

Closing of an open file is done by the command:

$$shut \ \ <user \ file>$$

or

$$shut \ \ "<user \ file>"$$

When all open files are closed, the output is automatically redirected to the terminal. An open file is generally lost if the current REDUCE session is terminated before it was shut, and an old file that is re-opened is totally erased before new output is written on it.

Normally, the output which is directed to a file is in the same form as it would appear on the terminal. In particular, integer powers of variables are printed with the exponents as superscripts, as is the natural writing style. However, if this file will be loaded later on in a REDUCE session, this form is not appropriate as REDUCE input. To suppress the raising of the exponents, the switch nat (which is normally on) can be put off. This is usually done before the file is opened. On closing the file and returning to terminal output, nat should be restored in its default status, on. As a side-effect, switching off nat will cause each output on the file to be terminated by a dollar sign, thus enabling further (silent) REDUCE interpretation of it.

In writing on a terminal or a file, REDUCE has a default number of characters on one line. To see the current maximum length of the line, use

$$linelength \ ()$$

To change this maximum length of the line, use the command

$$\text{linelength } <integer>$$

it returns the previous output linelength, so that it can be stored to reset it when necessary.

There is still another interesting file output format available, viz FORTRAN style output. This is useful if the results of the computations stored on a file are to be used in a FORTRAN program for further numerical calculations. Think e.g. of symbolic derivations of a function in combination with an iteration process to find the zeroes of that function. The command switch `fort`, when on, causes the output from that moment on to be FORTRAN-compatible. The number of continuation lines is limited by the value of the global variable `!*cardno`, whose default value is 20, but can be changed by the user if necessary. Statement values which are not assigned to an identifier are automatically directed to a FORTRAN variable named ans. This variable ans is also used as an intermediate variable to assign the piecewise evaluations of expressions exceeding the default number of continuation lines. Another name for this default variable can be fixed by the command

$$\text{varname } <identifier>$$

It is assumed that the results on the file will be used for floating-point calculations, so the integers in the expressions (but not the integer exponents) are written with a decimal point. However, if the results will be used in integer arithmetic, this decimal point can be suppressed by switching off `period`.

The numbers of characters on one line in FORTRAN style output is stored in the variable `fortwidth!*` and can be changed by the user. The actual linelength will be 6 less because FORTRAN style output starts at the seventh column.

Chapter 4

APPLICATIONS

4.1 A functional equation

Exponential functions can be characterized as the entire solutions to the functional equation

$$f(x + a) = bf(x), \qquad \forall x \in \mathbf{C}$$

(where a and b are constants), i.e. $f(x) = Cb^{x/a}$. We ask ourselves what happens to the solution if we replace the product on the right-hand side by an exponentiation. Because exponentiation is not commutative, there are at least two ways of doing this. One way leads to the equation

$$f(x + a) = f(x)^b.$$

Taking logarithms and writing $g(x) = \log f(x)$, we get $g(x + a) = bg(x)$, so $g(x)$ must be an exponential function and the solution of this equation is of the form $A^{b^{x/a}}$. The other introduction of an exponential leads to the equation

$$f(x + a) = b^{f(x)}.$$

Again, we stress that the function f must be analytic at least in an open set U such that U and $U + a$ have a non-empty intersection (this guarantees that the problem is not trivial). Performing a logarithmic transformation of the argument (i.e. introducing $g(\exp x) = f(x)$), we obtain the equation

$$g(cx) = b^{g(x)},$$

for which $x = 0$ is a special value, at which both sides express the behavior of g at the same point. This allows us to compute the derivatives of both sides at $x = 0$ and to equate them, yielding a set of conditions on g's Taylor coefficients.

For the zeroth derivative, we see that $g(0) = b^{g(0)}$, hence, introducing p_1 as a first parameter, such that $g(0) = p_1$, we can replace b by p_1^{1/p_1}. Next, the first derivative leads to

$$cg'(cx) = \frac{\log p_1}{p_1} p_1^{g(x)/p_1} g'(x),$$

which, at $x = 0$, gives
$$(c - \log p_1)g'(0) = 0.$$

If we assume that $g'(0)$ is non-zero, this equation fixes c as $\log p_1$ and leaves $g'(0)$ arbitrary. Such arbitrariness should not surprise us because if the functional equation holds for $g(x)$, it also holds for $g(lx)$, where l is an arbitrary constant. Hence we may put $g'(0) = 1$.

The computation of the higher derivatives rapidly becomes quite complicated. We can easily convince ourselves that each such derivation will yield an equation for the corresponding derivative of g, and that this equation will be linear in this derivative (but non-linear in the other ones). This computation of higher derivatives, and the solution of the equations, can be performed by the REDUCE algebraic mode program we will now describe:

```
operator fun;
equation:=exp(cc*fun(x,0)/exp cc)-fun(cc*x,0);
```

This is the functional equation. We have replaced $\log p_1$ by the constant cc, mainly for efficiency: it is best to minimize the number of operators used, so we prefer one exp to two logs. In the final results, this choice will prove useful too. Note also that we have replaced $f(x)$ by fun(x,0), i.e. the zeroth derivative of fun, or f. This is to allow the proper derivation of f(cc*x).

```
for all k,l let
    df(fun(x,k),x,l)=fun(x,k+1),
    df(fun(cc*x,k),x,l)=(cc)**l*fun(cc*x,k+1);
for all k let
    df(fun(x,k),x)=fun(x,k+1),
    df(fun(cc*x,k),x)=cc*fun(cc*x,k+1);
```

These two sets of for all rules merely express the relationship between fun(x,k) and derivation. Once they have been added to REDUCE, arbitrarily complex expressions involving them can be differentiated.

```
operator result;
```

We store the solutions obtained for fun(0,k) in the operator result. We could have used another structure instead (an array, for instance), but this would limit a priori the number of results that can be stored.

```
result(0):=exp cc;
result(1):=1;
```

We cannot solve the zeroth-order equation because it is not linear. We do not solve the first-order equation because it leads to an arbitrary value for result(1), which we prefer to replace by 1.

```
begin
    scalar eqnderived,eqnderived0;
    eqnderived:=df(equation,x);
    for i:=2:8 do
```

```
<<   eqnderived:=df(eqnderived,x);
     eqnderived0:=num sub(x=0,eqnderived);
     for j:=1:i do
         eqnderived0:=
             num sub(fun(0,i-j)=result(i-j),
                     eqnderived0);
     result(i):=rhs first solve(eqnderived0,fun(0,i));
     write "result(",i,")=",result(i)>>;
end;
```

This begin ... end block constitutes the imperative part of the code. We take successive derivatives of equation and store them in eqnderived (starting with the second derivative). The value of eqnderived at $x = 0$ is put into eqnderived0, and all occurrences of fun$(0,k)$ where k is smaller than the current order of derivation, are substituted by the previous results. (In fact, we only consider the numerator of eqnderived, because no derivative of f can occur in the denominator.) Then, we substitute the values of fun$(x,0)$ already known, and at each step only keep the numerator (since we are considering an equation of the form eqnderived0=0). The resulting linear equation is solved for fun$(0,i)$, whose value can be found as the right-hand side of the first (and only) element in the list returned by the solve operator.

Here are some of the results:

$$result(2) = \frac{CC}{CC} \\ E^{CC} * (CC - 1)$$

$$result(3) = \frac{CC^2 * (CC + 2)}{E^{2*CC} * (CC^3 - CC^2 - CC + 1)}$$

$$result(4) = \frac{CC^3 * (CC^3 + 5*CC^3 + 6*CC^2 + 6)}{E^{3*CC} * (CC^6 - CC^5 - CC^4 + CC^2 + CC - 1)}$$

We see that each result(n) is a product of $\exp(-(n + 1)cc)cc^{n-1}$ with a rational function in cc, of degree $n(n-1)/2$ in the denominator and degree $(n-1)(n-2)/2$ in the numerator. A closer look at the denominators (and using REDUCE's factorization feature) leads to their identification with the product of $\prod_{i=1}^{n-1}(cc^i - 1)$. The numerator seems harder to identify; we notice that the second coefficient is $n(n-1)/2 - 1$ and

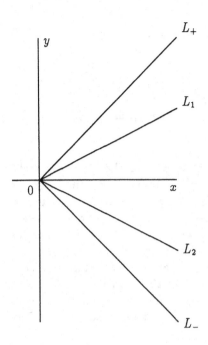

that the last coefficient is $(n-1)!$.

4.2   An algebra of one-dimensional projection operators

Recall the following fact from Euclidean geometry: given two n-dimensional subspaces A, B of some Euclidean space, their relative position is characterized by a set of principal angles $\{\theta_1, \ldots, \theta_n\}$, each in the interval $[0, \pi/2]$. In the following, P_X denotes the orthogonal projection on the vector subspace X.

There are different ways in which these angles can be found:

- the $\cos^2 \theta_i$ are the eigenvalues of $P_A P_B$ on A, or of $P_B P_A$ on B

- the $\sin^2 \theta_i$ are the eigenvalues of $(P_A - P_B)^2$, each appearing twice (if no θ_i is zero).

Example ——

Let L_1 and L_2 be lines in the plane (Cf. the figure).

The principal angle is the angle θ enclosed by L_1 and L_2. On L_1, $P_1 P_2$ is multiplication by $\cos^2 \theta$. Similarly, on L_2, $P_2 P_1$ is multiplication by $\cos^2 \theta$.

$P_1 - P_2$ has eigenvalue $\sin\theta$ on L_+ and $-\sin\theta$ on L_-. Its square $(P_1 - P_2)^2$ has eigenvalue $\sin^2\theta$ with the whole plane as eigenspace (so it has multiplicity 2).

Now consider the well-known problem of solving for a spherical triangle's angles when knowing its sides. Let A, B, C be its vertices, and write L_1 for the line joining 0 to A, L_2 for the line joining 0 to B and L_3 for the line joining 0 with C; our data (the angles a, b, c) then fix the relative position of each pair L_i, L_j, and we want to obtain the angle between two planes, say $\langle L_1 \cup L_2 \rangle$ and $\langle L_2 \cup L_3 \rangle$. Both 2-dimensional spaces have a nontrivial intersection (L_2), so one of their principal angles is zero; the other one is the angle we want to find.

We may also generalize these spherical triangles to hyperspherical simplexes and ask similar questions (e.g. what is the angle between $\langle L_1 \cup L_2 \cup L_3 \rangle$ and $\langle L_2 \cup L_3 \cup L_4 \rangle$, given the angles between each pair of lines).

To provide a general solution to this, we briefly study the algebra generated by the P_i (where P_i is the orthogonal projection on L_i), in order to implement the computations in this algebra using REDUCE. Note that we will work in general dimension, so we cannot just use arbitrary co-ordinates for the L_i and introduce the matrix representation of the P_i (that would be far too costly anyway).

Given $L_1, \ldots, L_k$, what is the structure of the linear transformation algebra generated by $\mathbf{R}$ (the products with a real scalar) and the P_i on $\langle L_1 \cup \ldots \cup L_k \rangle$? We choose unit vectors p_i on each L_i and write $\cos\theta_{ij} = p_i \cdot p_j$, where $\theta_{ij} \in [0, \pi]$.

In general, $P_1 P_2 \neq P_2 P_1$, so the algebra is not commutative. We prove the following lemmas:

- On $\langle L_1 \cup \ldots \cup L_k \rangle$, the following decomposition of the identity transformation 1 holds:

$$1 = \sum_{i,j=1}^{k} \frac{\Delta_{ij}}{\Delta} \cdot \frac{P_i P_j}{\cos\theta_{ij}}.$$

  Applied to a general element in the Euclidean space, the right-hand side yields its orthogonal projection on $\langle L_1 \cup \ldots \cup L_k \rangle$.

  (Here, if D represents the matrix $(\cos\theta_{ij})_{i,j=1}^{k}$, $\Delta = \det D$ and Δ_{ij} is the (i,j) cofactor of D, i.e. Δ_{ij}/Δ is the (j,i) element of D^{-1}.)

  Proof. Let x be arbitrary in $\langle L_1 \cup \ldots \cup L_k \rangle$ and decompose it using the affine basis $\{p_i\}$. Then

$$x = \sum_{i=1}^{k} x_i p_i.$$

Taking an inner product with p_j, $j = 1, \ldots, k$, we have

$$p_j \cdot x = \sum_{j=1}^{k} x_i (p_i \cdot p_j).$$

By Cramer's solution to a system of linear equations, x_i is therefore given by a quotient M_i/Δ, where M_i is the determinant of D where the ith column is replaced by the $p_j \cdot x$. Expanding this determinant along that column, we have:

$$x_i = \sum_{j=1}^{k} \frac{\Delta_{ij}}{\Delta}(p_j \cdot x)$$

and the result follows if we notice that $P_j x = (p_j \cdot x)p_j$, so that $P_i P_j(x) = (p_j \cdot p_i)(p_j \cdot x)p_i = \cos\theta_{ij}(p_j \cdot x)p_i$.   QED

- Let A be an arbitrary linear operator. Then for given i, j there exists a unique constant a such that $P_i A P_j = a P_i P_j$.

  Proof.

$$P_i A P_j x = (P_i A(p_j))(p_j \cdot x) = \frac{P_i A p_j}{p_i}(p_j \cdot x)p_i = \left(\frac{\frac{P_i A p_j}{p_i}}{\cos\theta_{ij}}\right)P_i P_j x.$$

                                                                        QED

- In particular,

$$P_i P_j P_k = \frac{\cos\theta_{ij}\cos\theta_{jk}}{\cos\theta_{ik}} P_i P_K$$

  Proof. Compute.   QED

Now we have the following: *the algebra generated by the P_i is the whole linear transformation algebra of $\langle L_1 \cup \ldots \cup L_k\rangle$ and any element in it can be uniquely written as a linear combination of the $P_i P_j$.*

Proof. Clearly, the algebra generated by the P_i consists of finite linear combinations of finite products of P_i. Any product of n, $(n > 2)$, P_i can be reduced to $n - 1$, then $n - 2$, etc. up to and including 2, P_i. A product of one P_i is equal to $P_i P_i$. A product of zero P_i is a scalar constant, which we can replace by a multiple of the decomposition of 1. Given any linear transformation A of $\langle L_1 \cup \ldots \cup L_k\rangle$, we have $A = 1 A 1$, and replacing the 1 here by their decomposition, we get a sum of terms of the form $P_i P_j A P_k P_l$, which reduce to scalar multiples of $P_i P_l$.   QED

So, to compute with the P_i and the θ_{ij}, it seems good to introduce a symbol for $P_i P_j$. Upon closer inspection, it is even better to introduce a symbol for $P_i P_j/\cos\theta_{ij}$. Call this $q(i,j)$. The corresponding REDUCE rules are then:

`operator p,q,c;`

$p(i)$ is the projection P_i, it simplifies to $q(i,i)$. $q(i,j)$ is $P_i P_j/\cos\theta_{ij}$. $c(i,j)$ is shorthand for $\cos\theta_{ij}$ (the cosines are not really necessary here).

`noncom p,q;`

Neither p or q are commutative; in REDUCE, non-commutativity is treated as a property, commutativity is default.

```
for all i let p(i)=q(i,i);
for all i,j,k,l let q(i,j)*q(k,l)=c(j,k)*q(i,l);
for all i,j let q(i,j)**2=c(i,j)*q(i,j);
```

Just compute to verify this. The third rule is required by the pattern matcher.

```
for all i,j such that i>j let c(i,j)=c(j,i);
```

This eliminates any redundancy due to the symmetry of c(i,j).

```
for all i let c(i,i)=1;
```

Indeed, $\theta_{ii} = 0$.

```
procedure trace e;
sub(q=c,e);
```

trace computes the trace of a simplified expression involving the q(i,j). Because trace $P_i P_j = \cos^2 \theta_{ij}$, trace q$(i,j)$ is c(i,j) and taking the trace just means replacing q by c. Of course, the simplifier will then eliminate the redundancy which is due to the symmetry of c, so e.g. q(1,2)-q(2,1) is non-zero, but its trace is c(1,2)-c(2,1), which simplifies to zero.

```
procedure proj l;
begin
    scalar n;
    n:=length l;
    matrix m(n,n);
    for i:=1:n do
        for j:=1:n do
            m(i,j):=c(part(l,i),part(l,j));
    m:=m**(-1);
    return
        for i:=1:n sum
            for j:=1:n sum
                m(i,j)*q(part(l,i),part(l,j));
end;
```

proj takes a list l of integer indices $i_1,\ldots$ and returns the orthogonal projection operator on the space generated by $L_{i_1},\ldots$, using the decomposition given above.

Examples ————————————————————————

```
p(1);
Q(1,1)
p(1)*p(2);
C(1,2)*Q(1,2)
trace p(1);
1
```

```
(P(1)-P(2))**2;
Q(2,2) - Q(2,1)*C(1,2) - Q(1,2)*C(1,2) + Q(1,1)
(1/2)*TRACE((P(1)-P(2))**2);
        2
 - C(1,2)  + 1
```

Now we can apply this to compute the angles of a spherical triangle, given its sides:

```
pa:=proj {1,2};
         - Q(2,2) + Q(2,1)*C(1,2) + Q(1,2)*C(1,2) - Q(1,1)
PA := -----------------------------------------------------
                            2
                      C(1,2)  - 1
pb:=proj {2,3};
         - Q(3,3) + Q(3,2)*C(2,3) + Q(2,3)*C(2,3) - Q(2,2)
PB := -----------------------------------------------------
                            2
                      C(2,3)  - 1
```

```
on gcd;  (1/2)*trace((PA-PB)**2);
        2                                      2          2
 - C(2,3)  + 2*C(2,3)*C(1,3)*C(1,2) - C(1,3)  - C(1,2)  + 1
------------------------------------------------------------
         2        2          2          2
   C(2,3) *C(1,2)  - C(2,3)  - C(1,2)  + 1
```

4.3   Gröbner bases

4.3.1   Systems of polynomial equations

In mathematical practice, many problems lead to systems of polynomial equations: examples are the computation of intersections of algebraic surfaces, and the determination of equations for certain geometric loci. Such a system is of the form

$$\begin{cases} p_1(z_1,\ldots,z_n) &= 0 \\ &\vdots \\ p_k(z_1,\ldots,z_n) &= 0 \end{cases},$$

where the p_j are elements of the ring $\mathcal{R}$ of n-variate polynomials with complex coefficients. We shall write P for the set $\{p_1,\ldots,p_k\}$ and Σ_P for the solution set of the system:

$$\Sigma_P = \{(z_1,\ldots,z_n) \in \mathbf{C}^n | \forall p \in P : p(z_1,\ldots,z_n) = 0\}.$$

There exist several methods to determine Σ_P from P:

- ad hoc combination of the polynomial equations to eliminate some of the variables or to substitute for them

- computing resultants (which may introduce parasitic solutions)

- determining a Gröbner base for *the ideal* generated by P.

In this section, we shall discuss the third method. For more details, we refer to e.g. [2].

4.3.2 Polynomial ideals

An essential difficulty is that P is not at all characteristic for Σ_P: for instance, if $k = 2$, $P = \{p_1, p_2\}$ can be replaced by $\{p_1 + p_2, p_1 - p_2\}$ (a non-singular linear transformation of the p_i), or by $\{p_1 - p_2^2 + p_1^2, p_2^2 - p_1^2\}$, or by many other, non-linear transformations. This is why we introduce a new concept, the ideal $I(P)$ generated by P:

$$I(P) = \{q_1 p_1 + \ldots + q_k p_k | q_j \in \mathcal{R}\}.$$

$I(P)$ is the set of all '$\mathcal{R}$-linear' combinations of the elements of P. In general, a subset I of $\mathcal{R}$ is called an *ideal* if

1. I is a ring

2. I is closed under multiplication by any element of $\mathcal{R}$, i.e. $I\mathcal{R} \subseteq \mathcal{R}$.

An ideal I gives rise to the following equivalence relation on $\mathcal{R}$: $f \sim_I g$ iff $f - g \in I$. To any ideal I of $\mathcal{R}$, one can also associate the maximal subset of $\mathbf{C}^n$ on which all elements of I vanish:

$$\Sigma_I = \{(z_1, \ldots, z_n) \in \mathbf{C}^n | \forall p \in I : p(z_1, \ldots, z_n) = 0\}.$$

Examples ──

- let I_1 be the ideal generated by $P_1 = \{z_1, z_2\}$, then $I_1 = z_1\mathcal{R} + z_2\mathcal{R}$ and $\Sigma_{I_1} = \{(z_1, \ldots, z_n) \in \mathbf{C}^n | z_1 = z_2 = 0\}$.

- Similarly, let I_2 be the ideal generated by $P_2 = \{z_1 + z_2\}$, then $I_2 = (z_1 + z_2)\mathcal{R}$ and $\Sigma_{I_2} = \{(z_1, \ldots, z_n) \in \mathbf{C}^n | z_1 + z_2 = 0\}$.

Note that $I_2 \subset I_1$ and that $\Sigma_{I_1} \subset \Sigma_{I_2}$: in general, the larger the ideal I, the smaller Σ_I. The largest ideal of $\mathcal{R}$ is $\mathcal{R}$ itself (it is generated by $\{1\}$); $\Sigma_{\mathcal{R}}$ is the empty set.

The following results are fundamental to polynomial ideal theory:

- if $n = 1$, every ideal I of $\mathcal{R}$ is generated by a single polynomial p, so that $I = p\mathcal{R} = I(\{p\})$;

- if $n > 1$, every ideal I of $\mathcal{R}$ is generated by a finite number of polynomials $p_1, \ldots, p_k$, so that $I = p_1\mathcal{R} + \ldots + p_k\mathcal{R} = I(\{p_1, \ldots, p_k\})$.

These generating polynomials p and p_j for I are not unique.

4.3.3   The ideal associated to a system of polynomial equations

Let P be a finite set of polynomials. It is easy to see that Σ_P and $\Sigma_{I(P)}$ coincide. This implies that a system of polynomial equations can always be viewed in terms of the ideal generated by them. On the other hand, because polynomial ideals can always be generated by a finite number of polynomials, each Σ_I is the solution set of a finite system of polynomial equations.

These facts apply to our problem as follows: given a system of polynomial equations P, we consider the ideal $I(P)$ generated by them and determine a 'better' set of generators P'—one for which the system P' is easier to solve.

4.3.4   Reduction of a polynomial

Given a polynomial f and a system P of polynomials, to reduce f with respect to P means replacing it by a polynomial g that is equivalent to f for the ideal $I(P)$ (i.e. $f - g \in I(P)$), but 'simpler' than f. This notion of 'simpler' has to be explained further—we distinguish two cases:

1. If $n = 1$, the monomials $1, z, z^2, \ldots$ are totally ordered by their degree, in a natural way. Because, in this case, the ideal $I(P)$ can be generated by a single nonvanishing polynomial p, and because there is a unique division property, viz that $f = pq + r$ for unique q, r satisfying $\deg r < \deg p$, we can interpret the property of being 'simpler' as 'having a lower degree'. The reduction of f with respect to P is then simply the polynomial rest r.

2. If $n > 1$, the monomials are no longer totally ordered by their degree (consider for instance z_1 and z_2). There is no unique division property by which to compute a reduction.

   We shall introduce a total ordering $\gg$ on the monomials, but cannot do so in a unique and natural way anymore. But we shall still require that the total ordering be preserved by multiplication, i.e.

   $$\text{for all monomials } a, b, c: \quad a \gg b \implies ac \gg bc.$$

   A total ordering satisfying this condition will be called *acceptable*.

The total orderings used in practice are based on some arbitrary ordering of the first degree monomials, the variables themselves. Usually, the variables will be listed in decreasing order, so that $\{x, y, z\}$ means that $x \gg y \gg z$ for the chosen total ordering. The product in a general monomial will also be written in decreasing order: $z_1^{d_1} \cdots z_n^{d_n}$ if $z_1 \gg z_2 \gg \cdots \gg z_n$.

Given two monomials with exponents $d = (d_1, \ldots, d_n)$ and $e = (e_1, \ldots, e_n)$, the preservation property motivates us to have $d \gg e$ if and only if $d - e \gg (0, \ldots, 0)$. There are three commonly used total orderings of this type:

(a) lex: the lexicographical ordering, for which $d \gg {}_{lex}(0)$ iff for some k, $d_1 = d_2 = \ldots = d_{k-1} = 0$ and $d_k > 0$

Example ────────────────────────────────────

If $x \gg y \gg z$, then $x^3 y z \gg x y^2 z^3$, because the degree in x is higher; $x y^3 z \gg x y^2 z^4$ because the degree in x is equal and the degree in y is higher, and, generally,

$$x^2 \gg xy \gg xz \gg x \gg y^2 \gg yz \gg y \gg z^2 \gg z \gg 1.$$

────────────────────────────────────

(b) gradlex: the total degree lexicographic ordering, for which $d \gg {}_{gradlex}(0)$ iff $\deg d > 0$ or $\deg d = 0$ and $d \gg {}_{lex}(0)$

Example ────────────────────────────────────

$x y^2 z^3 \gg x^3 y z$, $x y^2 z^4 \gg x y^3 z$, $x^2 y z^2 \gg x^2 z^3$.

────────────────────────────────────

(c) revgradlex: the total degree, reverse lexicographic ordering, for which $d \gg {}_{revgradlex}(0)$ iff $\deg d > 0$ or $\deg d = 0$ and $(0) \gg {}_{lex} d$.

Example ────────────────────────────────────

$x^2 z^3 \gg x^2 y z^2$.

────────────────────────────────────

Remark

Most REDUCE procedures in the Gröbner package accept an optional argument holding the list of variables, to fix their ordering; if it is not supplied, the internal ordering of identifiers is used (it can be modified using the korder command).

When the total ordering of the monomials has been fixed, polynomials are usually written in descending order of monomials, and the first one is called the leading monomial.

The property of being 'simpler' is then expressed using the leading monomials: if the leading monomial of g is $\gg$ that of f, f is considered to be 'simpler'.

Given a finite set of polynomials P, an ordering of the variables occurring in them, and a total ordering of the monomials, $f \in \mathcal{R}$ is said to be *reduced* with respect to P if its leading monomial is not a multiple of the leading monomial of some element of P.

If a polynomial f is not reduced with respect to P, a multiple of some element of P can be subtracted from f in order to eliminate its leading monomial; the resulting polynomial g will then have a leading monomial that is $\ll$ than the leading monomial of f. Furthermore, g is equivalent to f with respect to $I(P)$. This procedure is called the reduction of f with respect to P (it depends on the total ordering chosen).

Note that the reduction of f with respect to P may well be achieved in different ways, depending on which elements of P are chosen at each step. A single reduction step will usually not be sufficient to reduce f; however, a finite number of such steps will always suffice.

4.3.5   Complete reduction of a polynomial

A polynomial f is said to be *completely reduced* with respect to P if none of its monomials is a multiple of the leading monomial of some element of P.

Example ——

Let $P = \{g\} = \{y - 1\}$, $x \gg y$ and let the lexicographical ordering be chosen. Then $f = x + y^2 + y$ is reduced with respect to P, but it is not completely reduced, for y^2 and y occur in f. Subtracting yg from f gives $x + 2y$, and subtracting $2g$ then gives $x + 2$, which is completely reduced with respect to P.

The complete reduction of f with respect to P can be carried out in REDUCE using the procedure

$$\text{preduce}(f, \{P_1, \ldots, P_k\}).$$

Example ——

```
f:=x^2*y^2+x+y^5+y-7; u:=x^2+y^2-2; v:=x*y-1;
preduce(f,{u,v});
      5    4       2
X + Y  - Y  + 2*Y  + Y - 7
preduce(f,{v,u});
      5
X + Y  + Y - 6
```

```
korder y,x;
preduce(f,{u,v});
      4    3     2
5*Y - X  + X  + 2*X  - 3*X - 7
preduce(f,{v,u});
4*Y - X - 6
```

4.3.6 Gröbner bases

Given an ordering of the variables, an acceptable total ordering of the monomials and an ideal I in $\mathcal{R}$, a *standard* (or *Gröbner*[1]) base of I is a generating set of polynomials $\tilde{P}$ for which the reduction of any element of I yields zero. An equivalent definition is: ... such that every polynomial in $\mathcal{R}$ has a unique reduction with respect to $\tilde{P}$. Note that reduction is used in these definitions, not complete reduction.

Example ───
The set $P = \{u, v\}$ of the preceding example clearly is not a standard base.

───

It is not obvious whether an ideal necessarily possesses a standard base, nor how to construct one if possible. The first question is answered positively by one of the forthcoming theorems. It is remarkable that there exists an algorithm (due to Buchberger) to construct a standard basis given a set of generating polynomials for the ideal, an ordering of the variables and an acceptable total ordering of the monomials. This algorithm relies on a practical definition of Gröbner bases, given further on.

For a given ideal, a standard base is not unique. As one of the theorems indicates, a unique standard base can only be obtained under the additional condition of reducedness.

Theorem: every ideal I of $\mathcal{R}$ has at least one standard base with respect to each acceptable ordering.

Definition: a reduced standard base of an ideal is a standard base such that each of its elements is completely reduced with respect to the other elements.

Theorem: two ideals are equal if and only if they have the same reduced standard bases (for the same, given, acceptable ordering).

This is the canonical representation theorem for ideals with an acceptable ordering of monomials.

In REDUCE, the reduced Gröbner base of a given system $P = \{P_1, \ldots, P_k\}$ can be computed using the procedure call

$$\text{groebner}(\{P_1, \ldots, P_k\}\, [, \{v_1, \ldots, v_n\}]\,);$$

───

[1]Wolfgang Gröbner, 1899–1980.

Example ──

```
u:=x^2+y^2-2;  v:=x*y-1;
groebner({u,v},{x,y});
       3     4      2
{X  +  Y  - 2*Y,Y   - 2*Y   + 1}
groebner({u,v},{y,x});
       3     4      2
{Y  +  X  - 2*X,X   - 2*X   + 1}
```

──

The solution set of P is therefore $\Sigma_P = \{(-1,-1),(1,1)\}$; it consists of two points. Note that only one variable occurred in the last element of the Gröbner base. This is an instance of a more general fact:

Theorem: a system P of polynomial equations has a finite number of solutions (in complex space) if and only if each variable only occurs in one of the leading monomials of the associated Gröbner base for the lexicographical ordering.

In this case, one of the polynomials of the Gröbner base will only contain a single variable; it can be solved for the values of this variable. These can then be substituted in the other polynomial equations, one of which will again contain only one of the variables, etc. This way, all solutions can be recovered (if one computes in suitable extension fields when necessary).

4.3.7   Buchberger's algorithm

Given an ordering of the variables, an acceptable ordering of the monomials, and two polynomials f and g, we define as follows a polynomial linear combination $S(f,g)$ of f and g: let f_1, g_1 be the leading monomials of f and g and let h be their least common multiple, then

$$S(f,g) = \frac{h}{f_1}f - \frac{h}{g_1}g.$$

Note that $S(f,f) = 0$ and that $S(f,g) = -S(g,f)$.

Theorem: a generating set P of the ideal I is a standard base of I iff for all f, g in P, $S(f,g)$ can be reduced to 0 with respect to P.

Example ──

Let u and v be as before, then

$$S(u,v) = yu - xv = y^3 - 2y + x$$

which reduces to $x + y^3 - 2y$, a nonzero result, so that $\{u,v\}$ is not a Gröbner base. But if we consider $f = x^3 + y^3 - 2y$ and $g = y^4 - 2y^2 + 1$,

$$S(f,g) = y^4 f - xg = 2xy^2 - x + y^7 - 2y^5,$$

this reduces to 0.

──

The theorem indicates how a standard base can be computed from a given set of generators P for an ideal I: assume that P is not a standard base, then some $S(f,g)$ does not reduce to zero, but to some polynomial q_1. This polynomial is still in I, so we can add it to P, obtaining $P_1 = P \cup \{q_1\}$, and still $I(P_1) = I$. Next we consider again all $S(f,g)$, f and g in P_1, and possibly add other polynomials to P_1. Buchberger has proved the nontrivial fact that this process must stop after a finite number of steps, so that it can be used to construct a standard base.

Example ──

Let P be $\{u,v\}$ as before, take $x \gg y$ and choose the lexicographical ordering. Then $S(u,v)$ reduces to $x + y^3 - 2y$, which we will denote by w. Adding w to P, we see that $u = xw + (2 - y^2)v$, so that u can be eliminated from P.
Next, $S(v,w) = -y^4 + 2y^2 - 1$, which we will denote by $-s$. Adding s to P, we see that $v = yw - s$, so that v can be eliminated from it.
Finally, $S(w,s)$ reduces to zero, so that $\{w,s\} = \{x + y^3 - 2y, y^4 - 2y^2 + 1\}$ is a standard base for the ideal.

4.3.8   The radical ideal

Gröbner bases can also be used to check whether a given polynomial vanishes over the whole of an algebraic set.
We again consider a system of polynomial equations and its associated set P. If we introduce the set

$$V(P) = \{p \in \mathcal{R} | \forall (z_1, \ldots, z_n) \in \Sigma_P : p(z_1, \ldots, z_n) = 0\},$$

it is easy to verify that $V(P)$ is an ideal; it consists of all polynomials that vanish on the whole of Σ_P. It should also be clear that $I(P) \subseteq V(P)$.
The relationship between $I(P)$ and $V(P)$ is clarified by the following result:

Theorem (Hilbert's Nullstellensatz): a polynomial p belongs to $V(P)$ iff for some $m \in \mathbf{N}$, p^m belongs to $I(P)$.

So $V(P)$ consists of the polynomials that are radicals (mth roots) of some element of $I(P)$; it is therefore called the *radical* ideal of $I(P)$.
To test whether a polynomial belongs to $V(P)$, given P, the following fact is used:

Lemma: a system of polynomial equations is inconsistent iff the corresponding standard base (with respect to any of the acceptable orderings) contains a constant. (It must then be $\{1\}$.)

Proof:

1. Let the system corresponding to P be inconsistent: Σ_P is empty, so $I(P) = \mathcal{R}$, whose standard base is $\{1\}$.

2. Let 1 be an element of the standard base: it generates $\mathcal{R}$, so that $I(P) = \mathcal{R}$ and Σ_P is empty, i.e. the system is inconsistent.

Now let p and P be given. To determine whether p belongs to $V(P)$, we introduce a new variable, say ζ, and add $1 - \zeta p = 0$ to the system of equations, obtaining the set P^*.

If $p \in V(P)$, p vanishes on Σ_P, so that P^* is inconsistent and its standard base is $\{1\}$. Conversely, if P^*'s standard base is $\{1\}$, P^* is inconsistent and p must vanish on the whole of Σ_P.

Example

Let us check that

$$x = \frac{1 - z^2}{1 + z^2}, \qquad y = \frac{2z}{1 + z^2}$$

defines points (x, y) on the unit circle: rewriting the equations in polynomial form, we must test whether $x^2 + y^2 - 1$ is a radical element of the ideal generated by $x(1 + z^2) - (1 - z^2)$ and $y(1 + z^2) - 2z$.

```
problem:={x*(1+z^2)-(1-z^2),y*(1+z^2)-2*z}$
groebner problem;
                 2
{X + Y*Z - 1, Y*Z  + Y - 2*Z}
```

Now check whether $x^2 + y^2 - 1$ is a radical element:

```
equation:=x**2+y**2-1$
groebner((1-u*equation).problem);

{1}
```

It is. In fact, we see that

```
groebner(equation.problem);
                 2
{X + Y*Z - 1, Y*Z  + Y - 2*Z}
```

so that the equation is already an ordinary element of the ideal. Explicitly, we have that

$$x^2 + y^2 - 1 = (x + yz - 1)(x - yz + 1) + (yz^2 + y - z)y.$$

Now assume that we would be unenlightened on the geometric meaning of our equations for x and y, and that we would have started with the squares of the equations:

```
p2:={first(problem)**2,second(problem)**2};

       2 4      2 2    2      4         4        2
P2 := {X *Z  + 2*X *Z  + X  + 2*X*Z  - 2*X + Z  - 2*Z  + 1,


       2 4      2 2    2        3            2
      Y *Z  + 2*Y *Z  + Y  - 4*Y*Z  - 4*Y*Z + 4*Z }
```

The ideal $I(S)$ is different in this case:
groebner p2;

```
 2      2 2      2                        3 3    3          2 2     2
{X  - X*Y *Z  - X*Y  + 4*X*Y*Z - 2*X - Y *Z  - Y *Z + 4*Y *Z  + Y  - 4

 *Y*Z + 1,


 2 4       2 2    2         3                2
 Y *Z  + 2*Y *Z  + Y  - 4*Y*Z  - 4*Y*Z + 4*Z }
```

In fact, $x^2 + y^2 - 1$ is no longer an element of $I(S)$:
groebner(equation . p2);

```
 2    2
{X  + Y  - 1,


                2 3     2            2
 2*X*Y - 2*X*Z + Y *Z  + 3*Y *Z - 4*Y*Z  - 2*Y + 2*Z,


  2               2
 X*Z  - X - 2*Y*Z + Z  + 1,


 2 4       2 2    2         3                2
 Y *Z  + 2*Y *Z  + Y  - 4*Y*Z  - 4*Y*Z + 4*Z }
```

One can verify similarly that equation**2 is not an element of the ideal. From the decomposition given above, one can immediately infer that equation**3 will be an element, so equation is a radical element. This can also be verified directly:
*groebner((1-u*equation).p2);*

```
{1}
```

4.3.9   Euler's nine-point circle

Let ABC be a triangle. Then the following nine points lie on a circle:

- the midpoints K, L, M of the sides

- the feet P, Q, R of the altitudes

- the Euler points D, E, F: the midpoints of the segments joining the orthocenter H to the vertices.

Let $O(x, y)$ denote the center of the nine-point circle. Then the following equations suffice to determine x, y and the radius, ρ:

$$
\begin{aligned}
0 = g_1 &= (x - b - c)^2 + y^2 - \rho^2 && \text{the circle passes through } K \\
0 = g_2 &= (x - c)^2 + (y - a)^2 - \rho^2 && \text{the circle passes through } L \\
0 = g_3 &= (x - b)^2 + (y - a)^2 - \rho^2 && \text{the circle passes through } M
\end{aligned}
$$

To check whether P lies on the circle, we must verify that $x^2 + y^2 - \rho^2$ is a radical element of the ideal generated by g_1, g_2 and g_3:

```
problem:={g1,g2,g3}$
p1:=x^2+y^2-rho^2;
groebner((1-u*p1).problem,{u,x,y,z});
{1}
```

so this is verified.

Next, Q lies on AC and BQ is perpendicular to AC, i.e. if we denote its co-ordinates by (x_2, y_2),

$$
\begin{aligned}
0 = g_4 &= ax_2 + cy_2 - 2ac \\
0 = g_5 &= 2c(x_2 - 2b) - 2ay_2
\end{aligned}
$$

To check whether Q lies on the circle, we must verify that $p_2 = (x_2 - x)^2 + (y_2 - y)^2 - \rho^2$ is a radical element of the ideal generated by $\{g_1, g_2, g_3, g_4, g_5\}$:

```
problem:={g1,g2,g3,g4,g5}$
p2:=(x2-x)^2+(y2-y)^2-rho^2;
groebner((1-u*p2).problem,{u,x,y,z,x2,y2});
{1}
```

To verify that R, H, E and F also lie on the circle is done in the same way.

A convenient system of equations for the co-ordinates (x, y) of O and the radius ρ can be obtained as follows:

```
groebner({g1,g2,g3},{x,y,rho});
          B + C
{ X - -------,
          2

              2
          BC - A
  Y + ---------,
          2 A
             2 2     2 2    4      2 2
    2     B A  + B C  + A  + A C
  RHO  - ------------------------ }
                    2
                 4 A
```

Chapter 5

A PACKAGE FOR THREE-DIMENSIONAL EUCLIDEAN GEOMETRY

We present a set of procedures covering most classical constructs in three-dimensional euclidean geometry.

5.1 Vectors

The basic object type in this program is a vector: a linear combination of the three unit vectors e_1, e_2, e_3 that form an orthonormal right-handed basis.

A vector takes the form: $a_1*e(1) + a_2*e(2) + a_3*e(3)$ where the a_j, $j \in \{1,2,3\}$ are scalar expressions.

The unit vectors e(j) are defined by means of an operator, e, which is declared non-commutative for reasons that will be explained soon:

```
operator e;
noncom e;
```

Extracting a cartesian co-ordinate out of a vector can be done using comp:

```
procedure comp(j,u); df(u,e(j));
```

Next to addition and scalar multiplication, the fundamental operations on vectors include the scalar (or dot) product and the vector (or cross) product. These are defined by:

```
procedure dot(u,v);
u inner v;
```

```
procedure ext(u,v);
-(u outer v)*e(1)*e(2)*e(3);
```

Example ————————————————————————————————————

```
operator a,b;
vector1:=for j:=1:3 sum a(j)*e(j);
VECTOR1 := A(3)*E(3) + A(2)*E(2) + A(1)*E(1)
```

```
vector2:=for j:=1:3 sum b(j)*e(j);
VECTOR2 := B(3)*E(3) + B(2)*E(2) + B(1)*E(1)
dot(vector1,vector2);
A(3)*B(3) + A(2)*B(2) + A(1)*B(1)
ext(vector1,vector2);
 - (E(3)*(A(2)*B(1) - A(1)*B(2)) + E(2)*( - A(3)*B(1) + A(1)*B(3)) +
E(1)*(A(3)*B(2) - A(2)*B(3)))
```

These definitions of the dot and cross products rely on to two other operations, viz the inner and outer product of vectors, which are operations defined in a larger vector space, which is even an algebra, and which is treated of in the next section.

5.2   Clifford algebra

The *Clifford* (or *geometric*) algebra constructed over the three-dimensional euclidean space with orthonormal basis (e_1, e_2, e_3), is an eight-dimensional vector space generated by the basis

$$(1, e_1, e_2, e_3, e_2 e_3, e_3 e_1, e_1 e_2, e_1 e_2 e_3),$$

which is turned into a non-commutative algebra by introducing the 'geometric' product defined by the rules $e_j^2 = 1$ when $j \in \{1, 2, 3\}$ and $e_i e_j = -e_j e_i$ when $i, j \in \{1, 2, 3\}$ and $i \neq j$. In REDUCE:

```
for all j let e(j)^2=1;
for all j,k such that j>k let e(j)*e(k)=-e(k)*e(j);
factor e;
```

Examples ————————————————————————————————

```
vector1*vector2;
 E(2)*E(3)*( - A(3)*B(2) + A(2)*B(3))
 + E(1)*E(3)*( - A(3)*B(1) + A(1)*B(3))
 + E(1)*E(2)*( - A(2)*B(1) + A(1)*B(2))
 + A(3)*B(3) + A(2)*B(2) + A(1)*B(1)
vector1^2;
     2       2       2
A(3)  + A(2)  + A(1)
(e(2)*e(3))^2;
-1
(e(1)*e(2)*e(3))^2;
-1
```

```
(e(1)+e(1)*e(2))*(e(1)-e(1)*e(2));
 - (2*E(2) - 2)
(3*e(1)+e(1)*e(2))^2;
8
(1+e(1))*(1-e(1));
0
```
Notice the existence of zero-divisors!

In this algebra, there are four grades:

- the subspace spanned by 1 is the space of scalars: it can be identified with the space of the reals, **R**

- the subspace spanned by e_1, e_2, e_3 is the space of vectors, which is identified with the underlying euclidean space

- the subspace spanned by e_2e_3, e_3e_1, e_1e_2 is the space of bivectors

- the subspace spanned by $e_1e_2e_3$ is the space of trivectors or pseudoscalars.

The most general Clifford number can therefore be seen as the sum of a scalar, a vector , a bivector and a pseudoscalar.

The decomposition in these four grades can be obtained for by the procedure grade, which is defined via the auxiliary procedure gradsplit:

```
operator usergrad;
for all j,k let usergrad(j)*usergrad(k)=usergrad(j+k);
for all j let usergrad(j)^2=usergrad(2*j);

procedure gradsplit ex;
begin
    scalar y;
    y:=ex*usergrad(0);
    for j:=1:3 do
        y:=sub(e(j)=usergrad(1)*e(j),y);
    return y
end;

procedure grade(ex,k);
df(gradsplit ex,usergrad(k));
```

Examples ————————————————————————————————

```
x:=1+e(1)+e(1)*e(2)+e(1)*e(2)*e(3);
grade(x,0);
1
```

```
grade(x,1);
E(1)
grade(x,2);
E(1)*E(2)
grade(x,3);
E(1)*E(2)*E(3)
```

Inspection of the geometric product of two vectors (see the above example) leads to the introduction of the so-called inner and outer products of vectors and, more generally, of Clifford numbers. The inner product of two vectors is the scalar part of their geometric product (it is commutative), while their outer product is the bivector part of their geometric product (which is anti-commutative).

The general definitions are as follows:

```
procedure inner(a,b);
begin
    scalar as,bs,s,u;
    as:=gradsplit a;
    bs:=gradsplit b;
    s:=0;
    for j:=1:3 do
        << u:=df(as,usergrad(j));
            if u neq 0 then
                s:=s+( for k:=1:3 sum
                    grade(u*df(bs,usergrad(k)),abs(j-k)))>>;
    return s
end;


procedure outer(a,b);
begin
    scalar as,bs,s,u;
    as:=gradsplit a;
    bs:=gradsplit b;
    s:=0;
    for j:=0:3 do
        << u:=df(as,usergrad(j));
            if u neq 0 then
                s:=s+(for k:=0:3 sum
                    grade(u*df(bs,usergrad(k)),j+k))>>;
    return s
end;
```

Moreover, inner and outer are declared as infix operators and their precedence with respect to the other arithmetic operations is fixed:

```
infix inner;
infix outer;
precedence inner,*;
precedence outer,inner;
```

Examples _____

```
e(1) inner e(1);
1
e(1) outer e(1);
0
e(1) inner e(2);
0
e(1) outer e(2);
E(1)*E(2)
vector1 inner vector2;
A(3)*B(3) + A(2)*B(2) + A(1)*B(1)
vector1 outer vector2;
  - (E(2)*E(3)*(A(3)*B(2) - A(2)*B(3)) + E(1)*E(3)*(A(3)*B(1)
  - A(1)*B(3)) + E(1)*E(2)*(A(2)*B(1) - A(1)*B(2)))
```

The last two examples are in accordance with the above remark on the inner and outer products of two vectors. As to the inner and outer products of a vector and a bivector, they too arise from a partition of the geometric product according to the grade:

```
bivector1:=a(2,3)*e(2)*e(3)+a(3,1)*e(3)*e(1)+a(1,2)*e(1)*e(2);
BIVECTOR1 :=  - (- A(2,3)*(E(2)*E(3)) + A(3,1)*(E(1)*E(3))
              - A(1,2)*(E(1)*E(2)))
vector2*bivector1;
  E(3)*( - A(3,1)*B(1) + A(2,3)*B(2))
  + E(2)*( - A(2,3)*B(3) + A(1,2)*B(1))
  + E(1)*E(2)*E(3)*(A(3,1)*B(2) + A(2,3)*B(1) + A(1,2)*B(3))
  + E(1)*(A(3,1)*B(3) - A(1,2)*B(2))
```

This is the vector part of their product:

```
vector2 inner bivector1;
  E(3)*( - A(3,1)*B(1) + A(2,3)*B(2))
  + E(2)*( - A(2,3)*B(3) + A(1,2)*B(1))
  + E(1)*(A(3,1)*B(3) - A(1,2)*B(2))
```

This is the trivector part of their product:

```
vector2 outer bivector1;
E(1)*E(2)*E(3)*(A(3,1)*B(2) + A(2,3)*B(1) + A(1,2)*B(3))
```

Three (anti-)involutions are defined on the Clifford algebra; they are called reversion, main involution and conjugation. The corresponding procedures are rever, mainvol and bar.

The reversion is used to define to the so-called Clifford norm of a Clifford number cnorm, which need not to be scalar; its scalar part, however, coincides with the usual squared euclidean norm enorm.

Also notice the definition of the star product (the infix operator star) as the scalar part of the geometric product.

```
procedure rever ex;
begin
    scalar y;
    y:=gradsplit ex;
    y:=sub(usergrad(0)=1,usergrad(1)=1,
            usergrad(2)=-1,usergrad(3)=-1,y);
    return y
end;
```

```
procedure mainvol ex;
begin
    scalar y;
    y:=gradsplit ex;
    y:=sub(usergrad(0)=1,usergrad(1)=-1,
            usergrad(2)=1,usergrad(3)=-1,y);
    return y
end;
```

```
procedure bar ex;
begin
    scalar y;
    y:=gradsplit ex;
    y:=sub(usergrad(0)=1,usergrad(1)=-1,
            usergrad(2)=-1,usergrad(3)=1,y);
    return y
end;
```

It is readily seen that the conjugation (bar) is the composition of the reversion and the main involution.

```
procedure star(a,b);
grade(a*b,0);
```

```
procedure enorm ex;
star(ex,rever ex);

procedure cnorm ex;
ex*rever ex;

infix star;
precedence star,-;
```

Examples ───

```
x:=1+e(1)+e(1)*e(2)+e(1)*e(2)*e(3);
rever x;
  - (E(1)*E(2)*E(3) + E(1)*E(2) - E(1) - 1)
mainvol x;
  - (E(1)*E(2)*E(3) - E(1)*E(2) + E(1) - 1)
bar x;
E(1)*E(2)*E(3) - E(1)*E(2) - E(1) + 1
cnorm(e(1)+e(1)*e(2));
  - (2*E(2) - 2)
enorm(e(1)+e(1)*e(2));
2
```

───

We pointed out before that not every non-zero Clifford number has an inverse. However, if the Clifford norm of a Clifford number is positive, the inverse clearly exists; sadly, it is not a necessary condition, as is seen by the following example:

```
cnorm(3*e(1)+e(1)*e(2));
  - (6*E(2) - 10)
```

yet

```
(3*e(1)+e(1)*e(2))*(3*e(1)+e(1)*e(2))/8;
1
```

5.3 Groups in the Clifford algebra

The Clifford group Γ of a Clifford algebra $\mathcal{A}$ consists of those invertible elements whose action on a vector produces again a vector. This action of a Clifford number on another one is defined as follows:

```
procedure action(a,b);
bar(a)*b*a;
```

For elements of Γ, it is an orthogonal automorphism of the three dimensional euclidean space.

The positive numbers constitute a normal subgroup of Γ, and the quotient group is called the Pin group of the Clifford algebra.

The subalgebra of the Clifford algebra $\mathcal{A}$ consisting of linear combinations of scalars and bivectors is called the even subalgebra $\mathcal{A}^+$; the subspace consisting of linear combinations of vectors and pseudoscalars is called the odd subspace $\mathcal{A}^-$.

The intersection of the even subalgebra $\mathcal{A}^+$ and Γ is a subgroup Γ^+; its quotient over the positive reals is called the Spin group. The intersection of the odd $\mathcal{A}^-$ subspace with Γ yields the set Γ^-, and its intersection with the Pin group yields the OPin set. In the three-dimensional case, Γ^+ holds exactly the nonzero elements of the even subalgebra; in higher dimensions, matters are more complicated.

Notice that the even subalgebra $\mathcal{A}^+$ is isomorphic to the algebra of *quaternions*; it is sufficient to make the following identifications:

$$i \leftrightarrow e_2 e_3, \quad j \leftrightarrow e_3 e_1 = -e_1 e_3, \quad k \leftrightarrow e_2 e_1 = -e_1 e_2.$$

Given an arbitrary Clifford number, the procedure group will determine to which group it belongs. It uses a lot of auxiliary predicates.

```
% a predicate that tests whether its argument is a list
lisp operator listp;
lisp procedure listp x;
(pairp x) and (car x = 'list);
```

To exclude lists as arguments to the following boolean procedures, we qualify any expression which is not a list, as a Clifford expression (without considering matrices, arrays and the like). This is tested by the following predicate:

```
procedure clifp x;
if listp x then nil else t;


procedure realp x;
if clifp x and x=grade(x,0) then t else nil;


procedure vectorp x;
if clifp x and x=grade(x,1) then t else nil;


procedure bivectorp x;
if clifp x and x=grade(x,2) then t else nil;


procedure trivectorp x;
if clifp x and x=grade(x,3) then t else nil;


procedure aplusp x;
if clifp x and x=grade(x,0)+grade(x,2) then t else nil;
```

```
procedure aminp x;
if clifp x and x=grade(x,1)+grade(x,3) then t else nil;

procedure gammaplusp x;
if aplusp x and x neq 0 then t else nil;

procedure gammaminp x;
if aminp x and x neq 0 then t else nil;

procedure spinp x;
if aplusp x and cnorm x = 1 then t else nil;

procedure opinp x;
if aminp x and cnorm x =1 then t else nil;

procedure gammap x;
if gammaplusp x or gammaminp x then t else nil;

procedure pinp x;
if spinp x or opinp x then t else nil;

procedure group x;
begin scalar v,r;
    if not clifp x then
        <<  write "this is not a Clifford number";
            return nil>>
    else if x=0 then
        <<  write "this is zero!";
            return nil>>;
    r:=rever x;
    v:=x*r;
    if not realp v then
        <<  write "this Clifford number is not in the Clifford
                    group, however, its inverse might exist";
            return nil>>
    else if not vectorp(action(x,! x)) then
        <<  write "this Clifford number is not in the Clifford
                group, however it is invertible with x^(-1) = ",r/v;
            return nil>>
    else if aplusp x then
        <<  if v = 1 then
```

```
                    <<  write "this Clifford number is in the spingroup";
                        return nil>>
                    else write "this Clifford number is in the group "
                                "gammaplus and ",
                                x/sqrt(v)," is in the spingroup ";
                    return nil>>
            else if aminp x then
                <<  if v = 1 then
                    <<  write "this Clifford number is in the pin group",
                            "more precisely in the opin subspace";
                        return nil>>
                    else write "this Clifford number is in the group gamma",
                                "more precisely in the gammamin subspace, ",
                                x/sqrt(v)," is in the pingroup",
                                "more precisely in the opin subspace";
                    return nil>>
end;
```

Examples ——

group(3*e(1)+e(1)*e(2));

```
this Clifford number is not in the Clifford group, however
its inverse might exist
```

Remark: since $(3e_1 + e_1 e_2)^2 = 8$, we know that this Clifford number is indeed invertible!

group(1+e(1)*e(2)*e(3));

```
this Clifford number is not in the Clifford group, however
it is invertible with x^(-1) =
                    E(1)*E(2)*E(3) - 1
                 - ---------------------
                             2
```

group(1+e(1)*e(2));

```
this Clifford number is in the group gammaplus and
                    E(1)*E(2) + 1
                    ---------------
                      .SQRT(2)
is in the spingroup
```

group(e(1)+e(1)*e(2)*e(3));

```
this Clifford number is in the group gamma, more precisely
in the gammamin subspace,
```

$$\frac{E(1)*E(2)*E(3) + E(1)}{SQRT(2)}$$

is in the pingroup, more precisely in the opin subspace

5.4   The geometric meaning of elements of the Clifford group

Elements of the Clifford group have a specific geometric meaning:

- the action of a vector is a reflection with respect to the plane through the origin perpendicular to that vector

- a Spin element being the product of two unit vectors, its action is the composition of two reflections, a rotation.

- an element of OPin being the product of at most three unit vectors, its action is the composition of a reflection and a rotation.

Given a Clifford number of the Clifford group, the procedure geom will determine the geometric meaning of the action of that number. If the given Clifford number does not belong to the Clifford group, a warning results. If the Clifford number is in the Clifford group but not in the Pin group, it is first normalized, i.e. divided by its norm.

```
procedure geom x;
if not gammap x then
    << write "this clifford number is not in the clifford group";
       nil>>
else if spinp x then
    geomspin x
else if opinp x then
    geomopin x
else
    << write "your clifford number is normalized";
       geom(x/sqrt(x*rever x)) >>;

procedure geomspin x;
begin
    scalar a,b,c;
    a:=grade(x,2);
    if a=0 then
        << write "the action of this spin element 1 is ",
                 "the identity transformation";
```

```
            return nil >>;
    c:=x-a;
    b:=sqrt(cnorm a);
    a:=-a*e(1)*e(2)*e(3)/b;
    write "the action of this spin element is ",
          "the rotation about the axis ",
          a,
          " with rotation angle 2*theta, where theta is ",
          "given by cos(theta) = ",c,
          " and by sin(theta) = ",b;
    return nil
end;


procedure geomopin x;
begin
    scalar a,b,aa,bb;
    a:=grade(x,1);
    b:=x-a;
    if a=0 then
        <<  write "the action of this opin element is ",
                  "the reflection w.r.t. the origin";
            return nil >>
    else if b=0 then
        <<  write "the action of this vector is ",
                  "the reflection w.r.t. the plane through ",
                  "the origin, perpendicular to that vector";
            return nil >>;
    aa:=sqrt(cnorm a);
    bb:=-b*e(1)*e(2)*e(3);
    write "the action of this opin element is the rotation ",
          "about the axis ",-a/aa,
          " with rotation angle 2*theta where theta is ",
          "given by cos(theta) = ",bb,
          " and by sin(theta) = ",aa,
          " followed by the reflection w.r.t. the origin";
    return nil
end;
```

Examples ——

```
geom 0;
this Clifford number is not in the Clifford group
geom 7;
```

your Clifford number is normalized
the action of this spin element 1 is the identity transformation
geom(1+e(1)*e(2));

your Clifford number is normalized
the action of this spin element is the rotation about the axis E(3)
with rotation angle 2*theta, where theta is given by

```
                  1                                1
cos(theta) =  --------- and by sin(theta) = ---------
                SQRT(2)                          SQRT(2)
```

geom e(2);

the action of this vector is the reflection
with respect to the plane through the origin,
perpendicular to that vector

geom(e(1)+e(1)*e(2)*e(3));

your Clifford number is normalized
the action of this opin element is the rotation about the axis - E(1)
with rotation angle 2*theta where theta is given by cos(theta) =

```
    1                                1
--------- and by sin(theta) = ---------
 SQRT(2)                          SQRT(2)
```

followed by the reflection w.r.t. the origin

geom(e(1)*e(2)*e(3));

the action of this opin element is the reflection w.r.t. the origin

---

Conversely, if a rotation axis and angle are given, the corresponding Spin element
is given by rotator(*axis, angle*); similarly, given a plane, the corresponding OPin
element is given by reflector(*plane*). Here *axis* stands for a list of two vectors and
plane for a list of three vectors; the representation of lines and planes by means of
lists of vectors is explained in the next section on objects.

```
procedure rotator(axis,angle);
begin
    scalar u;
    u:=second axis - first axis;
    u:=u/sqrt(cnorm u);
    return cos(angle/2)
           +sin(angle/2)*u*e(1)*e(2)*e(3)
end;

procedure reflector(plane);
begin
```

```
    scalar u,a,b,c;
    a:=first plane;
    b:=second plane;
    c:=third plane;
    u:=ext((b-a),(c-a));
    return u/sqrt(cnorm u)
end;
```

Examples ——————————————————————————————

```
13:={0,e(3)}; rotator(13,pi/2);
 SQRT(2)*(E(1)*E(2)) + SQRT(2)
-------------------------------
              2
```

which is verified by
geom ws;
the action of this spin element is the rotation about the axis E(3)
with rotation angle 2*theta, where theta is given by cos(theta) =
```
 SQRT(2)                                 1
--------- and by sin(theta) = ---------
    2                                 SQRT(2)
v:={e(1),e(2),e(3)}; reflector v;
 E(3) + E(2) + E(1)
--------------------
        SQRT(3)
```

which is verified by
geom ws;
the action of this vector is the reflection w.r.t. the plane
through the origin, perpendicular to that vector

——————————————————————————————————————

This representation of rotations and reflections by Clifford numbers is particularly
suited for computing the composition of such orthogonal transformations.

Example ——————————————————————————————

```
axis1:={0,e(1)-e(2)+e(3)}; angle1:=2*pi/3;
axis2:={0,e(2)-e(3)}; angle2:=pi;
rota1:=rotator(axis1,angle1);
          E(2)*E(3) + E(1)*E(3) + E(1)*E(2) + 1
ROTA1 := -----------------------------------------
                            2
```

rota2:=rotator(axis2,angle2);

$$\text{ROTA2} := - \frac{E(1)*E(3) + E(1)*E(2)}{\text{SQRT}(2)}$$

The composition of the two rotations...
rota:= rota1*rota2;

$$\text{ROTA} := - \frac{E(1)*E(2) - 1}{\text{SQRT}(2)}$$

...is again a rotation.
geom rota;

```
the action of this spin element is the rotation about the axis  - E(3)
with rotation angle 2*theta, where theta is given by cos(theta) =
     1                                      1
--------- and by sin(theta) = ----------
 SQRT(2)                           SQRT(2)
```

A perhaps more familiar look at those orthogonal transformations in terms of matrices can be obtained by means of the operator tfmat. For a given Pin element x, tfmat(x) constructs the corresponding orthogonal matrix m and returns the value of its determinant; the matrix m then is simply asked for by the command m;.

```
procedure tfmat(g);
begin
    if not pinp g then
        <<  write "this clifford number is not in the pin group";
            return nil >>;
    matrix m(3,3);
    for k:=1:3 do
        <<  for j:=1:3 do
                m(j,k):=DF(action(g,e(k)),e(j)) >>;
    return det m
end;
```

5.5   Objects

5.5.1

The geometrical objects we are dealing with are points, lines, planes and the trivial object of space itself. Each of these is represented by a list of vectors—of so-called point vectors, bound vectors connecting the special point chosen as the origin with a point in space. The number of such vectors in the list determines the nature of the object: a point is a list of one vector, a line a list of two vectors and a plane a list of three vectors.

An element of a list is lifted out as follows:

```
symbolic operator entry;
symbolic procedure entry(j,1);
nth(1,j+1);
```

The boolean function objectp checks whether a given argument is a geometrical object, i.e. a list of vectors:

```
procedure objectp lv;
begin
    scalar m;
    if not listp lv then
        return nil
    else if lv={} then
        return nil
    else
        m:=lv;
    while m neq {} and vectorp(first m) do
        m:=rest m;
    if m={} then
        return t
    else
        return nil
end;
```

This predicate is paraphrased by the following interrogation procedure, which provides a yes-or-no answer instead of t or nil.

```
procedure object!? lv;
if objectp lv then
    write "yes"
else
    write "no";
```

A fundamental role in this program is played by the procedure normal which checks a given list for superfluous vectors and converts it into a list of minimal length representing the same geometrical object.

```
procedure normal lv;
begin
    scalar elem,m;
    if lv={} then
        return {}
    else if not objectp lv then
        << write "this is not an object";
            return nil>>
```

```
        else if length lv < 2 then
            return lv;
        elem:=first lv;
        m:=normal(rest lv);
        if pincip({elem},m) then
            return m
        else
            return elem.m
end;
```

Examples ——

```
normal {e(1),e(1)};
{E(1)}
normal {(e(1)+e(2))/2,e(1),e(2)};
{E(1),E(2)}
normal {(e(1)+e(2)+e(3))/3,e(1),e(2),e(3)};
{E(1),E(2),E(3)}
```

5.5.2

Given a list of vectors, the procedure what determines its nature and returns respectively the co-ordinates of a point, the equations of a line and the equation of a plane, with respect to the orthogonal frame based on the orthonormal basis (e_1, e_2, e_3).

If the nature of an object is known, the procedures whatpoint, whatline and whatplane result into a list containing respectively the co-ordinates of the point, the equations of the line and the equation of the plane.

```
procedure whatpoint(p);
{comp(1,first p),comp(2,first p),comp(3,first p)};

procedure whatline(p);
begin
    scalar g;
    g:=ext(! x-first p,second p - first p);
    return {comp(1,g),comp(2,g),comp(3,g)}
end;

procedure whatplane(p);
{   det( mat ( (x1,x2,x3,1),
    (comp(1,first p),comp(2,first p),comp(3,first p),1),
    (comp(1,second p),comp(2,second p),comp(3,second p),1),
```

```
    (comp(1,third p),comp(2,third p),comp(3,third p),1)))};

procedure what(pp);
begin
    scalar p,l;
    if pp={} then
        <<  write "the Void";
            return nil>>
    else if not listp pp then
        <<  write "the argument should be an object or a figure";
            return nil>>
    else if objectp pp then
        <<  p:=normal pp;
            l:=length p;
            if l = 0 then write "the Void"
            else if l = 1 then
                write "the point with co-ordinates ",whatpoint(p)
            else if l = 2 then
                write "the line with equations ", whatline p
            else if l = 3 then
                write "the plane with equation ", whatplane p
            else if l = 4 then write "space">>
    else if figurep pp then
        <<  p:=figuredif pp;
            for each object in p do what object>>
    else write "nonsense";
end;
```

Examples ───

Consider the following points, lines and planes be given (Cf. figure).

```
o:={0};
p1:={e(1)}; p2:={e(2)}; p3:={e(3)};
r1:={e(2),e(3)}; r2:={e(3),e(1)}; r3:={e(1),e(2)};
l1:={0,e(1)}; l2:={0,e(2)}; l3:={0,e(3)};
v1:={0,e(2),e(3)}; v2:={0,e(3),e(1)}; v3:={0,e(1),e(2)};
v:={e(1),e(2),e(3)};
tetra:={0,e(1),e(2),e(3)};

what 0;
the argument should be an object or a figure
what o;
the point with co-ordinates {0,0,0}
```

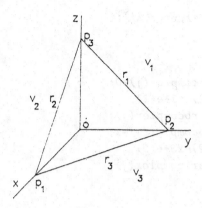

```
what p1;
the point with co-ordinates {1,0,0}
what r1;
the line with equations {X2 + X3 - 1, - X1, - X1}
what v;
the plane with equation {X1 + X2 + X3 - 1}
what tetra;
space
```

5.5.3

Conversely, if a point is given by a list of its co-ordinates, a line by a list of two equations, or a plane by a list holding its equation, the procedures point, line and plane respectively construct the corresponding geometrical objects as lists of vectors.

```
procedure point(l);
{first l*e(1)+second l*e(2)+third l*e(3)};

procedure line(ll);
begin
    scalar u,v,l,p,q,w,x1,x2,x3;
    u:=first ll;
    v:=second ll;
    l:=u/v;
    if df(l,x)=0 and df(l,y)=0 and df(l,z)=0 then
        << write "the equations are linearly dependent";
```

```
            return {}>>
    else w:=solve({u,v},{x,y,z});
    x1:=sub(w,x);
    x2:=sub(w,y);
    x3:=sub(w,z);
    for all j let arbcomplex(j)=0;
    p:=x1*e(1)+x2*e(2)+x3*e(3);
    for all j clear arbcomplex(j);
    for all j let arbcomplex(j)=1;
    q:=x1*e(1)+x2*e(2)+x3*e(3);
    for all j clear arbcomplex(j);
    return {p,q}
end;


procedure plane(u);
begin
    scalar w,p,q,r,x1,x2,x3;
    w:= solve(u,{x,y,z});
    x1:=sub(w,x);
    x2:=sub(w,y);
    x3:=sub(w,z);
    for all j let arbcomplex(j)=0;
    p:=x1*e(1)+x2*e(2)+x3*e(3);
    for all j clear arbcomplex(j);
    for all j let arbcomplex(j)=j;
    q:=x1*e(1)+x2*e(2)+x3*e(3);
    for all j clear arbcomplex(j);
    for all j let arbcomplex(j)=j+1;
    r:=x1*e(1)+x2*e(2)+x3*e(3);
    for all j clear arbcomplex(j);
    return {p,q,r}
end;
```

Examples ──

```
point {a,b,c};
{C*E(3) + B*E(2) + A*E(1)}
line {a1*x+b1*y+c1z+d1, a2*x+b2*y+c2*z+d2};
{
   E(2)*(A1*D2 - A2*C1Z - A2*D1) + E(1)*( - B1*D2 + B2*C1Z + B2*D1)
 - -------------------------------------------------------------,
                       A1*B2 - A2*B1
```

```
(E(3)*(A1*B2 - A2*B1) + E(2)*( - A1*C2 - A1*D2 + A2*C1Z + A2*D1) +
  E(1)*(B1*C2 + B1*D2 - B2*C1Z - B2*D1))/(A1*B2 - A2*B1)}
```

plane {a*x+b*y+c*z+d};

```
      D*E(1)
{ - --------,
       A


  3*A*E(3) + 2*A*E(2) + E(1)*( - 2*B - 3*C - D)
  ---------------------------------------------,
                       A


  4*A*E(3) + 3*A*E(2) + E(1)*( - 3*B - 4*C - D)
  ---------------------------------------------}
                       A
```

5.6   Incidence and coincidence of objects

Given two geometrical objects, the boolean function incip determines whether or
not the first object is incident or coincides with the second one—more explicitly: it
returns true if two points coincide, if a point lies on a line or in a plane, if two lines
coincide, if a line lies in a plane or if two planes coincide. If the first argument is an
empty list or the second one is space itself, it also returns trivially to true.
Given two objects <*object 1*> and <*object 2*>, the simple question inci!?(<*object
1*>,<*object 2*>) returns a yes or no answer depending on the incidence of both
objects as explained above.

Examples ———————————————————————————————

(with the same objects as before)
incip(p1,p1);

T

inci!?(r1,v2);

no

inci!?(r1,v);

yes

The boolean procedure objectequalp determines whether or not two given geometrical objects are equal (which might be interesting since the representation of each
geometrical object is far from unambiguous).

Example ──

```
l1:={0,e(1)};
objectequalp(l1,{-e(1),5*e(1)});
T
```

──

5.7 Union and intersection of objects

Given two geometrical objects p and q, we can already decide (by means of the boolean functions described in the preceding section) if they are incident or not.
Their intersection can be obtained by the procedure inter. There is also the possibility for obtaining their 'union' using the procedure uni. This notion of 'union' has to be understood in the following way: the 'union' of two different points is a line, the 'union' of a line and a point not on that line or the 'union' of two intersecting or parallel lines is a plane, etc.

Examples ──

```
line:=uni(o,p1);
LINE := {0,E(1)}
plane:=uni(line,l2);
PLANE := {E(1),0,E(2)}
uni(plane,l3);
{E(1),E(2),0,E(3)}
what ws;
space
inter(v,o);
{}
what ws;
the Void
inter(v,l1);
{E(1)}
what ws;
the point with co-ordinates {1,0,0}
inter(l1,r1);
the two lines are crossing

{}
```

```
inter(v,v2);
{E(3),E(1)}
what ws;
the line with equations

 - X2 = 0

X1 + X3 - 1 = 0

 - X2 = 0
```

5.8   Orthogonality of objects

5.8.1

Given a point and another geometrical object, the procedure **perpline** constructs the perpendicular through the point on the object, provided of course that the latter is a line or a plane.

Examples ————————————————————————————————
(still with the same objects as before)
```
perpline(o,v);
      E(3) + E(2) + E(1)
{0,--------------------}
             3
perpline(o,r1);
      E(3) + E(2)
{0,-------------}
         2
```
Notice that in both examples the perpendicular is given as the list consisting of the given point and the corresponding foot on the given object. Also notice the results of the following two examples:
```
l1:={0,e(1)}; perpline(o,l1);

indeterminate
```
(The given point is situated on the given line.)
```
v1:={0,e(2),e(3)}; perpline(o,v1);
{Y1*E(1),0}
```
(The result contains a parameter, and hence an arbitrary point of the perpendicular, since the given point is located in the plane.)

5.8.2

The procedure perpplane allows to construct the plane going through a point or a line which is perpendicular to a line or a plane; both arguments of this procedure thus are objects which have to match geometrically, otherwise a warning message results.

Examples ——

```
perpplane(p1,11);
{E(1),E(3) + E(1), - (E(2) - E(1))}
what ws;
the plane with equation
X1 - 1 = 0
perpplane(r1,11);
{E(2),E(3),E(3) + 2*E(2)}
what ws;
the plane with equation
 - 2*X1 = 0
perpplane(13,v);
     E(3) + E(2) + E(1)
{0,--------------------,E(3)}
              3
what ws;
the plane with equation
 X1 - X2
--------- = 0
    3
perpplane(13,v3);
indeterminate since the line is perpendicular to the plane
{}
perpplane(13,r1);
impossible
```

——

5.8.3

The boolean function ortop determines whether or not two given geometrical objects are orthogonal; the procedure orto!? reformulates the answer in 'yes' or 'no' terms.

Example ———
```
orto!?(v,{0,e(3),e(1)+e(2)+e(3)});
yes
```

5.8.4

Given two crossing lines, the procedure comperp constructs their common perpendicular.

Example ────────────────────────────────

```
l1:={0,e(1)}; r1:={e(2),e(3)};
comperp(l1,r1);
```

```
the two lines are crossing
   E(3) + E(2)
{-------------,0}
      2
```

5.9   Distance and objects

Given two geometrical objects, the procedure dist computes their mutual distance, which of course can be zero (if the objects are incident).

Examples ────────────────────────────────

(with the same objects as before)
```
dist(o,r1);
   1
---------
 SQRT(2)
```
```
dist(l1,l2);
0
```
```
dist(l1,r1);
the two lines are crossing
    1
---------
 SQRT(2)
```
```
dist(o,v);
   1
---------
 SQRT(3)
```

Using the distance function dist, a unit vector with the direction of a given line can now be computed. This is done by the procedure unitvector. If the argument happens to be of another nature than a line, the message 'impossible' appears.

Example ──

```
line:=perpline(o,v);
              E(3) + E(2) + E(1)
LINE := {0,--------------------}
                     3

unitvector line;
 SQRT(3)*E(3) + SQRT(3)*E(2) + SQRT(3)*E(1)
-------------------------------------------
                    3
```

──

5.10   Parallelism of objects

5.10.1

Given two geometrical objects, the procedure para constructs a third geometrical object which contains the first one and is parallel to the second, on condition that such a construction is possible and unambiguous; otherwise, an 'impossible' message is printed.

Examples ──

```
(with the same objects as before)
l:=para(p2,l1);
L := {E(2),E(2) + E(1)}
w:=para(o,v);
W := {0,E(2) - E(1),E(3) - E(1)}
para(r1,r1);
the two lines coincide
indeterminate since both lines coincide
ww:=para(l1,r1);
the two lines cross
WW := {0,E(1),E(3) - E(2)}
```

──

5.10.2

The boolean function parap determines whether or not two given objects are parallel; the analogous result in terms of 'yes' and 'no' is obtained by the command para!?.

Examples ——

para!?(l,11);

the two lines are parallel
yes

para!?(ww,r1);

the line is parallel to the plane
yes

——

5.11   Orthogonal transformation of objects

If a geometrical object is subjected to a rotation, with a given axis and rotation angle, or to a reflection with respect to a plane, its new position in space can be computed using the procedures objectrotation (applied to (<*axis*>,<*angle*>,<*object*>)) and objectreflection (applied to (<*plane*>,<*object*>)).

Examples ——

objectrotation(13,pi/3,r2);

```
        SQRT(3)*E(2) + E(1)
{E(3),--------------------}
                 2
```

objectreflection(v,o);

```
  2*E(3) + 2*E(2) + 2*E(1)
{-------------------------}
             3
```

——

5.12   Figures

Finally, we deal with figures: these are sets of objects that constitute some construct in space. For example, a cube may be viewed as a figure consisting of 8 specific points, or as a figure consisting of 6 specific planes, or as a figure consisting of 12 specific lines, or as a mixture of these. A figure is represented by a list of objects.

5.12.1

We first introduce a series of boolean functions which determine whether a given list represents a figure, a list of points, or lines or planes; they are called figurep, pointlistp, linelistp and planelistp, respectively.

Examples ────────────────────────────────────

cube:={o,p1,p2,p3,{e(1)+e(3)},{e(1)+e(2)},{e(2)+e(3)}, {e(1)+e(2)+e(3)}};

figure!? cube;

yes

pointlistp cube;

T

linelistp cube;

(no answer)

──

5.12.2

The boolean functions pointmemberp, linememberp, planememberp, objectmemberp and omemberp check if a given point or line or plane belongs to a given list of similar objects, and if a given object belongs to a given figure:

Examples ────────────────────────────────────

objectmemberp(p1,cube);

T

objectmemberp(l1,cube);

(no answer)

──

5.12.3

Given a figure, one can wonder if the objects in its representing list are all different. The procedure figuredif turns a figure-list into a figure-list that represents the same figure, but only contains pairwise different objects. Analogous procedures exist for figures consisting of objects that are all of the same nature, i.e. lists of points, of lines or of planes. All these procedures use the boolean functions introduced above to determine the geometrical character of given lists. Also the procedure what, already encountered when discussing objects, applies to figures and displays the contents of the representing figure-list.

Examples ───

```
figure1:={o,o,p1,p1,p2,p3};
figure:=figuredif(figure1);
FIGURE := {{0},

         {E(1)},

         {E(2)},

         {E(3)}}
what ws;
the point with co-ordinates {0,0,0}
the point with co-ordinates {1,0,0}
the point with co-ordinates {0,1,0}
the point with co-ordinates {0,0,1}
```

───

5.12.4

Given two figures, the boolean function figurequalp determines whether both figures are the same, i.e. consist of the same objects.

Example ───

(with the same figure and figure1 as above):
```
figurequalp(figure,figure1);
T
```

───

5.12.5

Given a figure consisting of points only, the procedure centroid determines—as its name indicates—the centroid of this set of points. To know whether those points are collinear or coplanar, use the boolean functions collinearp or coplanarpointp respectively, or ask the questions collinear!? or coplanarpoint!?, respectively.

Examples ───

```
p:=centroid(cube);
      E(3) + E(2) + E(1)
P := {--------------------}
             2
```

```
collinear!? cube;
```
no
```
coplanarpoint!? cube;
```
no

5.12.6

Given a figure consisting of only lines or only planes respectively, the boolean functions concurp, coplanarlinep, fanp and sheafp determine whether the lines are concurrent or coplanar, or whether the planes form a fan or a sheaf.

Examples ───────────────────────────────────────
```
fig1:={l1,l2,l3}; fig2:={r1,r2,r3}; fig3:={v1,v2,v3};
fig4:={v1,v2,{0,e(3),e(1)+e(2)}};
concur!? fig1;
```
yes
```
concur!? fig2;
```
no
```
concur!? fig3;
```
this is not a list of lines
no
```
coplanarline!? fig1;
```
no
```
fan!? fig2;
```
this is not a list of planes
no
```
fan!? fig3;
```
no
```
fan!? fig4;
```
yes
```
sheaf!? fig3;
```
yes
```
sheaf!? fig4;
```
no

5.12.7

Figures can also be subjected to rotations and reflections. The appropriate commands are

$$\texttt{figurerotation}(<axis>,<angle>,<figure>)$$

and

$$\texttt{figurereflection}(<plane>,<figure>).$$

Examples ──
```
axis:={0,e(1)+e(2)+e(3)}; angle:=2*pi/3;
figurerotation(axis,angle,cube);
 {{0}, {E(2)}, {E(3)}, {E(1)}, {E(2) + E(1)}, {E(3) + E(2)},
 {E(3) + E(1)}, {E(3) + E(2) + E(1)}}
figureequalp(ws,cube)
T

plane:={e(3)/2,e(3)/2 + e(1), e(3)/2 + e(2)};
figurereflection(plane,cube);
{{E(3)}, {E(3) + E(1)}, {E(3) + E(2)}, {0}, {E(1)},
 {E(3) + E(2) + E(1)}, {E(2)}, {E(2) + E(1)}}
figureequalp(ws,cube)
T
```
──

5.13 Conditions

In many geometrical problems, objects that contain parameters in their definition, have to fulfill certain conditions concerning incidence, orthogonality, parallelism etc. These conditions yield equations involving the parameters as unknowns.

The following procedures express the geometrical conditions by producing equations. Throughout this series of procedures, it is assumed that the argument objects are of the prescribed type; no normalization of the arguments takes place and the geometric object type is not checked. The user should supply the correct geometrical objects as arguments to these procedures, otherwise meaningless condition systems may result.

Given two points, the procedure cond2point expresses the condition under which both points coincide.

Example ──
```
p:=point {a,b,c}; q:=point {aa,bb,cc}; cond2point(p,q);
{A - AA = 0, B - BB = 0, C - CC = 0}
```

Given three points, the procedure cond3pointline expresses the condition under which the three points are collinear.

Example ————————————————————————————————————

cond3pointline(o,p,q);

{ B*CC - BB*C=0, -A*CC + AA*C = 0, A*BB - AA*B=0 }

As o represents the origin, the co-ordinates of the vectors p and q should be linearly dependent.

———

Given a point and a line, the procedure condpointline expresses the condition under which the point is situated on the line.

Example ————————————————————————————————————

condpointline(p,l1);

{0=0, - C=0,B=0}

Remember that l1 is the x-axis.

———

Given four points, the procedure cond4pointplane expresses the condition under which the four points are coplanar.

Example ————————————————————————————————————

cond4pointplane(o,p2,p,q);

{ - A*CC + AA*C=0}

Remember that (o, p2) is the y-axis, so the second co-ordinates of the points p and q are not involved.

———

Given a point and a plane, the procedure condpointplane expresses the condition under which the point is located in the plane.

Example ————————————————————————————————————

condpointplane(p1,v);

{0=0}

This condition is trivially fulfilled.

———

Given two lines, the procedure cond2line expresses the condition under which the two lines coincide.

Example ————————————————————————————————————

cond2line(l1,r1);

{1=0,0=0,0=0,1=0,1=0,1=0}

Clearly impossible.

Given two lines, the procedure `condlineinterline` expresses the condition under which both lines are intersecting.

Example ──

`r:=uni(p,q); condlineinterline(r,12);`

the two lines are crossing

`{A*CC - AA*C=0}`

See the above example with cond4pointplane.

Given three lines, the procedure `cond3lineplane` expresses the condition under which the three lines are coplanar.

Example ──

`cond3lineplane(11,12,r);`

`{0=0, - C=0, - CC=0}`

Naturally since 11 and 12 span the (x, y)-plane.

Given a line and a plane, the procedure `condlineplane` expresses the condition under which the line lies entirely in the plane.

Example ──

`condlineplane(r,v);`

`{ - A - B - C + 1=0, - AA - BB - CC + 1=0}`

Naturally since the equation of v reads $x + y + z = 1$.

Given two planes, the procedure `cond2plane` expresses the condition under which both planes coincide.

Example ──

`vv:=uni(r3,{d*e(3)}); cond2plane(vv,v3);`

`{0=0,0=0, - D=0}`

Indeed, since v3 is the (x, y)-plane.

Given three planes, the procedure `cond3planeline` expresses the condition under which the three planes have a line in common.

Example ————————————————————————————————

```
cond3planeline(vv,v,v3);
{0=0,0=0}
```

——

Given three planes, the procedure cond3planepoint expresses the condition under which the three planes have a point in common.

Example ————————————————————————————————

```
u:=para({d*e(3)},v3); cond3planepoint(v2,v3,u);
the line is parallel to the plane

the two lines are parallel

impossible, unless both lines coincide for which
                        the condition reads

{0=0,

    2
 - D  + D=0,

 0=0,

 0=0,

    2
 - D  + D=0,

 0=0}
```

Indeed the three planes only have a point in common if $D = 0$, i..e. if v3 and u, which are parallel, coincide.

——

Given two lines, the procedure condorto2line expresses the condition under which the lines are orthogonal.

Example ————————————————————————————————

```
op:=uni(o,p); oq:=uni(o,q); condorto2line(op,oq);
{A*AA + B*BB + C*CC=0}
```

——

Given a line and a plane, the procedure condortolineplane expresses the condition under which the line is perpendicular to the plane.

Example ——
condortolineplane(op,v3);
{B=0, - A + B=0}
which means that p should lie on the *z*-axis.
——

Given two planes, the procedure condorto2plane expresses the condition under which both planes are orthogonal.

Example ——
opq:=uni(op,q); condorto2plane(opq,v3);
{A*BB - AA*B=0}
which indeed means that the orthogonal projections on the plane v3 of both lines op and oq have to coincide.
——

Given two lines, the procedure condpara2line expresses the condition under which both lines are parallel.

Example ——
condpara2line(op,oq);
{B*CC - BB*C=0,

 - A*CC + AA*C=0,

 A*BB - AA*B=0}
which indeed means that both lines have to coincide.
——

Given two planes, the procedure condpara2plane expresses the condition under which both planes are parallel.

Example ——
condpara2plane(v1,v2);
{0=0,0=0,1=0}
Indeed impossible.
——

Given a line and a plane, the procedure condparalineplane expresses the condition under which the line is parallel to the plane.

Example ——
condparalineplane(r,v3);
{ - C + CC=0}
——

5.14 Applications

We illustrate the use of this software by working out two geometrical problems.

5.14.1 Desargues's Theorem verified

Desargues's Theorem states that if two triangles are perspective (i.e., the lines connecting corresponding vertices are concurrent), the corresponding sides of the triangle meet at three collinear points.

We assume—without loss of generality—that the triangles a,b,c and aa,bb,cc are perspective from the origin o. In the transcript, the non-relevant REDUCE output is omitted.

`o:={0};`

We take three arbitrary points a,b,c in the (e_1, e_2) plane as the vertices of the first triangle:

`a:={a1*e(1)+a2*e(2)}; b:={b1*e(1)+b2*e(2)}; c:={c1*e(1)+c2*e(2)};`

On their projecting lines oa, ob, oc from the origin, arbitrary points aa,bb,cc are taken as the vertices of the second triangle.

`oa:=uni(o,a); ob:=uni(o,b); oc:=uni(o,c);`
`aa:={sub(lamba=para1,first arbpoint oc)};`
`bb:={sub(lamba=para2,first arbpoint ob)};`
`cc:={sub(lamba=para3,first arbpoint oc)};`

```
AA := { - (E(2)*C2*(PARA1 - 1) + E(1)*C1*(PARA1 - 1))}
BB := { - (E(2)*B2*(PARA2 - 1) + E(1)*B1*(PARA2 - 1))}
CC := { - (E(2)*C2*(PARA3 - 1) + E(1)*C1*(PARA3 - 1))}
```

The sides of the first triangle are given by
`ab:=uni(a,b); bc:=uni(b,c); ca:=uni(c,a);`

The corresponding sides of the second triangle are given by
`aabb:=uni(aa,bb); bbcc:=uni(bb,cc); ccaa:=uni(cc,aa);`

The three intersecting points of the corresponding sides then are
`p:=inter(ab,aabb); q:=inter(bc,bbcc); r:=inter(ca,ccaa);`

Now we determine the conditions under which those three points p,q,r are collinear:
`cond3pointline(p,q,r);`

The following result is returned:
`{0=0,0=0,0=0}`

meaning that there is no condition at all, i.e. the points p,q,r are collinear!

5.14.2 Problem 2

Given: the line A with equations $x - 2 = 2y = k(-z - 1)$ (k being a real parameter), the plane alfa with equation $x + 2y + z - 1 = 0$, and the plane beta with equation $x + my + z - 10 = 0$ (m being also a real parameter)...

In REDUCE terms:

```
u:=x-2-2*y;  v:=2*y+k*(z+1);  a:=line {u,v};
        K*E(2) + 2*E(1)*(K - 2)
A := { - -------------------------,
                   2

    E(3) - K*E(2) + 2*E(1)*( - K + 1)}
w1:=x+2*y+z-1; alfa:=plane {w1};
ALFA := {E(1),

        5*E(3) + 4*E(2) - 12*E(1),

        6*E(3) + 5*E(2) - 15*E(1)}
w2:=x+m*y+z-10; beta:=plane {w2};
BETA := {10*E(1),

        7*E(3) + 6*E(2) + 3*E(1)*( - 2*M + 1),

        8*E(3) + 7*E(2) + E(1)*( - 7*M + 2)}
```

...answer the following questions:

1. For which values of the parameter k is the line A parallel to the plane alfa?
   `condparalineplane(a,alfa);`

   `{2*K - 1=0}`

   so the answer is $k = 1/2$.

2. For which values of the parameters k and m is the line A perpendicular to
   the plane beta; also determine, for the obtained values of the parameters, the
   co-ordinates of the foot.

`condortolineplane(a,beta);`

```
  14*K*M + 9*K + 16     2*K*M + K + 2
{--------------------=0,---------------=0}
          2                   2
```

This non-linear system is easily handled by the groebner algorithm:
`groebner ws;`

```
      1
{M - ---,K + 1}
      2
```

so the answer is $k = -1$, $m = 1/2$.

`let k=-1; let m=1/2;`

We look for an arbitrary but fixed point on the line A:

`p:=arbpoint a;`

```
P := { -
```

$$2*E(3)*(LAMBA - 1) + E(2)*(LAMBA - 2) + 2*E(1)*(LAMBA - 4)$$

```
------------------------------------------------------------}
                              2
```

We fix that point p by giving the parameter `lamba` a specific value:

`lamba:=4;`

`LAMBA := 4`

The perpendicular from that point p onto the plane beta is given by:

`perpline(p,beta);`

`{ - 3*E(3) - E(2),3*E(3) + 2*E(2) + 6*E(1)}`

We know that, in this representation of the perpendicular, the first point is p, while the second one is the foot, say q:

`q:={second ws 36};`

`Q := {3*E(3) + 2*E(2) + 6*E(1)}`

We check that q is indeed a point of the plane beta:

`incip(q,beta);`

`T`

The answer is therefore

`what q;`

`the point with co-ordinates {6,2,3}`

Bibliography

[1] A.G. Akritas, *Elements of Computer Algebra, with applications*, Wiley-Interscience (1989).

[2] B. Buchberger, G.E. Collins, R. Loos, R. Albrecht (eds.), *Computer Algebra—Symbolic and Algebraic Computation (2nd ed.)*, Springer-Verlag (1982).

[3] J.H. Davenport, Y. Siret, E. Tournier, *Computer Algebra, systems and algorithms for algebraic computation*, Academic Press (1988).

[4] A.C. Hearn, *REDUCE User's Manual (Version 3.4)*, RAND Publication CP78 (Rev. 7/91) (1991).

[5] M.A.H. MacCallum, F.J. Wright, *Algebraic Computing with REDUCE*, Clarendon Press (to appear).

[6] J.B. Marti, A.C. Hearn, M.L. Griss, C. Griss, *The Standard LISP Report*, University of Utah (1978).

[7] G. Rayna, *REDUCE: Software for Algebraic Computation*, Springer-Verlag (1987).

[8] G.L. Steele, *Common LISP: the language*, Digital Press (1984).

[9] J.A. van Hulzen, *Formulemanipulatie m.b.v. REDUCE*, Universiteit Twente (1989).

257

INDEX